science³

A Science Student's Success Guide

FIONA RAWLE
University of Toronto at Mississauga

TODD NICKLE
Mount Royal University

ROBERT THACKER
Saint Mary's University

SUSAN MORANTE
Mount Royal University

NELSON / EDUCATION

NELSON / EDUCATION

Science³: A Science Student's Success Guide

by Fiona Rawle, Todd Nickle, Robert Thacker, and Susan Morante

Vice President, Editorial Higher Education:
Anne Williams

Publisher:
Paul Fam

Marketing Manager:
Leanne Newell

Developmental Editor:
Candace Morrison

Photo Researcher:
Melody Tolson

Permissions Coordinator:
Melody Tolson

Content Production Manager:
Christine Gilbert

Production Service:
Integra Software Services Pvt. Ltd.

Copy Editor:
Julia Cochrane

Proofreader:
Erin Moore

Indexer:
Integra Software Services Pvt. Ltd.

Senior Manufacturing Coordinator:
Ferial Suleman

Design Director:
Ken Phipps

Managing Designer:
Franca Amore

Interior Design:
Dianna Little

Interior images:
Chessmen silhouettes, chess pieces: Francesco Abrignani/Shutterstock; Black minimalistic science icon set: liskus/Shutterstock

Cover Design:
Dianna Little

Cover Image:
Transgenic mouse embryo: Gloria Kwon; methane ice: *Energy from Fiery Ice, http://www.nrc-cnrc .gc.ca/eng/achievements/highlights/ 2008/energy_fiery_ice.html,* Natural Research Council of Canada, 2008. Reproduced with the permission of the Minister of Public Works and Government Services Canada, 2013; spiral galaxy somewhere in deep space: Vadim Sadovski/Shutterstock; Planet Earth with sunrise in space (main image): Alan Uster/ Shutterstock.

Compositor:
Integra Software Services Pvt. Ltd.

Printer:
RR Donnelley

Library and Archives Canada Cataloguing in Publication Data

Science³ : a science student's success guide / Fiona Rawle ... [et al.]. – 1st ed.

Includes bibliographical references and index.
ISBN 978-0-17-666273-8

1. Science—Handbooks, manuals, etc. I. Rawle, Fiona, 1978- II. Title: Science 3. III. Title: Science three. IV. Title: Sciencethree.

Q158.5.S35 2013
500   C2012-908602-9

PKG ISBN-13: 978-0-17-666273-8
PKG ISBN-10: 0-17-666273-1

BRIEF

Table of Contents

Contents

Contents

Contents

Contents

Contents

Contents

Contents

Contents

Contents

NOTE TO THE STUDENT

Welcome. This book is your guide to finding everything you need to know to study science. The information found within these pages is information that you usually won't get in a university classroom. It tells you how physics, chemistry, and biology inter-relate, and shows you how you can effectively apply the same study strategies to all three disciplines. This book was created by four authors from different disciplines of science, all of whom have learned that the typical student response of "siloing" information—clumping information into separate packages (silos) as they study—has hindered students' development as scientists. Although we specialize in physics or chemistry or biology, we recognize that *science* is our profession; we are simply interested in different aspects of the natural world.

We decided to write this book for a few reasons. For one, we think a guide that deals with *science* is overdue. Most student supplements cater to physics *or* chemistry *or* biology. Many students who are undertaking a science degree at a university will take a number of different science discipline courses outside their preferred field, particularly in the first couple of years. This is excellent exposure, and the breadth of our approaches to different facets of science should inspire you to think differently about the world around you. Scientists who make the most impact in their field are usually very well rounded; they may be inspired by other scientists—some work outside of their preferred field, many are interested in art and music, and a great number of them spend time on hobbies and projects that simply amuse them. Richard Feynman, a Nobel laureate in physics, loved to play the bongo drums and spent a year investigating the crumpling characteristics of tinfoil—simply because he loved to think about anything and everything.

Although there are four authors, we thought that writing this book as a collective, using phrases such as "we think" or "we suggest" in the chapters, would force a distance between us and you. It sounds authoritarian. Although we have slightly different emphases in our work, we all agree that we respect our students and value our roles as mentors to students who wish to excel at science. The tone we decided upon was to use the first person.

As we wrote, we imagined having an actual conversation as if between a mentor and a student. We want to give advice but not be prescriptive, and phrasing our ideas as "I think you should consider…" seems more personal. Our desire is to close the gap and provide insight and advice that we ourselves would have valued when we were undergraduates.

In these pages you'll find that the nature of science demands slightly different perspectives. While examples are drawn from the different fields of science, we try to show where they intersect as well as providing tables and lists that are tailored to help you out when you're actually working with biology, chemistry, or physics subjects. Despite the different perspectives, there are some very strong commonalities, particularly when you're trying to learn something new. An important section of this book goes into the best way to study for post-secondary classes and points out common myths and dead-end strategies that a lot of students employ. Many students view university as a place to learn lots of facts. It's not. Although you'll have to learn some things, being able to put them together or deconstruct ideas—to demonstrate actual *understanding* of complex things—is the key gateway to higher levels of study and research. And this takes time. There is no magic formula—no magic "download" strategy. Do you recall in the movie *The Matrix* when Neo (played by Keanu Reeves) asks Trinity (played by Carrie-Anne Moss), "Can you fly that thing?" Trinity responds by saying, "Not yet." She calls Tank (also known as the operator), and says, "I need a pilot program for a B2-12 helicopter." And voila—a pilot program gets downloaded into her brain and she responds to Neo, "Let's go." Many students think that science is like that: "If I memorize this textbook of facts, then I'll be able to do science and get an A." Even if it were possible to insert textbooks of knowledge into your brain, it wouldn't necessarily help you "do" science. As you'll learn in Chapter 2, science is a process, not just a collection of facts, and the emphasis of science education is more commonly on skill development, rather than on content regurgitation.

You'll also find a valuable section in this book that demonstrates and explains how to leverage the use of word roots, prefixes, and suffixes to save yourself time and energy when reading textbooks and scientific papers. Mathematics is a tool used by all scientists, so of course there's a chapter on that! Once you've learned all these new techniques and made discoveries, as a scientist you'll need to be able to communicate your ideas, so another chapter addresses this skill. There's even a chapter that shows you how to highlight your science skills to create better opportunities for a job or graduate school. This book can't possibly cover all the tiny details and formulas you'll require throughout all science courses, and it doesn't claim to. What this text provides is a consolidation of a variety of good starting points for your science education, with broad advice that will help you in a number of subjects.

NOTE TO THE EDUCATOR

This book should also provide some relief for educators. We've noticed in our institutions that professors and teaching assistants spend a lot of time going over writing requirements or defining units or giving study hints to students. The advice in this book and the resources for how to study, how to think critically, how science works, and how to communicate what you know can be tapped by instructors of introductory physics, chemistry, or biology. Students get better with practice. If you provide instruction and perspectives that are consistent between disciplines, your students will be able to focus more carefully on the science topics themselves because the underlying premises—how to write, what to know, how to think critically—will become an integral part of their thought processes as they improve their learning strategies.

A commitment to evidence-based pedagogy is one of the key goals of this book. However, in some chapters we cover topics that are, unavoidably, a matter of debate. Most notably, in our discussions of how science operates in Chapter 2, we have chosen to stay with the standard picture of Popperian falsification. It remains a good framework for beginning as well as experienced scientists regardless of its philosophical shortcomings.

TO ALL OUR READERS

We are very excited about this project. Although there's no way to make learning effortless, it's our wish that this handbook will show students how to *work smart* in addition to *working hard*. Of course, you must spend a lot of energy to become excellent at …

Preface

well … anything that's worthwhile. But it's our belief that considering the tips in this book, along with having the core information on a number of facets of science at your fingertips, will make for better students and give all of you—students and educators—a higher payoff for your efforts.

Fiona Rawle
Todd Nickle
Robert Thacker
Susan Morante
2013

ABOUT THE AUTHORS

Fiona Rawle, **Ph.D.**, has a doctorate in Pathology and Molecular Medicine and is currently a teaching-stream faculty member in the Department of Biology at the University of Toronto Mississauga. Fiona teaches Introduction to Evolution and Evolutionary Genetics, Introduction to Genetics, and Molecular Basis of Disease, and is actively involved in researching student learning about biology. The solutions to many of the crises that currently face human society will be based in science, from resource depletion to climate change to species extinction to rising cancer rates. The driving force behind Fiona's commitment to biology and to her students is that she is painfully aware of the impact these crises will have on our students (and future generations of students), and she sees the potential her students have to become the great scientists and problem solvers of the future. Fiona is also a TV Ontario (TVO) 2010 Best Lecturer Nominee. Here are some student comments from the Ontario educational broadcaster's website:

"She loves her subject, Biology, so much, and is so excited about teaching, her students can't help but be excited to come to class prepared and ready to learn."

"Dr. Fiona Rawle is not only knowledgeable about biology, but she is extremely passionate about it. When she conveys information, it is not all strictly memorization and facts, but it is put into the context of situations that apply to real life."

Todd Nickle, Ph.D., is a Professor in the Department of Biology at Mount Royal University. Todd graduated with a doctorate in Botany from Oklahoma State University in 1998 and since then has served in a faculty position at Mount Royal University in Calgary. He's been a leader and participant in teaching advocacy groups, such as the Alberta Introductory Biology Association and the Association for Biology Laboratory Education, and is ever eager to discuss teaching and learning strategies for students. His most recent project is to create teaching/learning resources that act as a "Professor in a Box," whereby he aims to bring biology to students when and where they want to access it. Having taught at universities for over 14 years, Todd recognizes the demands on students' time and has been experimenting with blended instructional delivery to improve performance (a preview of which was included in *Multimedia Modules to Accompany the Purple Pages,* in *Biology: Exploring the Diversity of Life,* First Canadian Edition [Nelson Education Ltd., 2010]). Todd's resources are intended to act as "training wheels," sometimes incorporating full-blown lectures delivered electronically at the students' convenience, helping to bolster their skills as they strive toward more technical and demanding academics.

Robert Thacker, Ph.D., is an Associate Professor, Canada Research Chair (Tier II), and Chairperson of the Department of Physics and Astronomy at Saint Mary's University (SMU). As well as being an award-winning and internationally renowned research specialist in cosmological large-scale structure and galaxy formation, he is also a passionate science communicator and appears regularly on local TV and radio. This keen interest in the public's understanding of science led Rob to investigate physics education research techniques in more detail, and he actively uses many of these techniques, particularly emotional involvement and the layering of concepts, in his teaching. Rob wants his students to be successful in today's increasingly competitive world, and feels an acute responsibility to provide information about the subject they're studying in a way that connects them directly to the material. His courses and evaluations consistently rank in the highest quintile at SMU. In the words of his students: "He puts emphasis not only on what is important, but on why it is important."

Preface

Susan Morante, M.Sc., is an Associate Professor and the Chair of the Department of Chemistry at Mount Royal University. Before beginning her career in post-secondary education, Susan worked in various research and industrial labs. During that time she did research on the red blood cells of premature infants, leading to a redesign of baby formula; set up protocols for intact, isolated cell membranes known as "ghosts"; and developed the formula for the first fluorescent highlighter pen to be patented in Canada. In addition to having almost 30 years' experience at Mount Royal University, she has taught at the University of Calgary. She specializes in the development and teaching of courses in chemistry, having taught most of the first- and second-year courses at one time or another. She currently teaches organic chemistry and is well loved by her students. She is also extremely interested in developing student research projects for third- and fourth-year chemistry students. To that end she has recently been working with her students on studying the pigments in bee hair and is really excited about a novel approach developed entirely by one of her students.

IN ADDITION TO THIS BOOK

Science is always changing as new perspectives are added and old ideas are challenged. Some of the conventions in physics, chemistry, and biology are changing as well. The *Science*[3] online site, science3.nelson.com, has the advantage of being updated on the fly to reflect current best practices. It can also be accessed from any place with an Internet connection, so you won't have to run to your bookshelf to find this book if you're in the library and want to check something. The site also provides multimedia so you can watch a demonstration of how to make a graph in Microsoft Excel (and, better yet, it covers both Macintosh and PC versions of Excel!).

Here are some of the features you'll find on the *Science*[3] website:

- Online support, examples, demonstrations, exercises, and troubleshooting tips.
- Sample essays in various formats. For example, one essay in each subject area will be shown in contrasting citation styles (i.e., APA and CSE) to highlight their differences.

- Word roots and how to use them to demystify the language of science.

- Video tutorials about how to set up and solve mathematics problems.

- Student activities and workbook components, including detailed step-by-step help for students.

ACKNOWLEDGMENTS

We would like to acknowledge the many people throughout the science community who have contributed ideas and information that have been incorporated into this book. Mindy Thuna, of the University of Toronto at Mississauga, authored Appendix A. The credit for using the scene from *The Matrix* as an example of how learning does not work goes to Professor Carol Rolheiser of the Centre for Teaching Support and Innovation at the University of Toronto.

We are indebted to the many experts in the sciences who carefully reviewed this book at several stages for their outstanding and invaluable advice on how to construct an effective textbook:

Edward Andrews, Grenfell Campus, Memorial University of Newfoundland
Wendy Benoit, University of Calgary
Jason Donev, University of Calgary
Lori Jones, University of Guelph
Benjamin Kelly, York University
Krystyna Koczanski, University of Manitoba
Derek Lawther, University of Prince Edward Island
Celeste Leander, University of British Columbia
Pippa Lock, McMaster University
Ben Newling, University of New Brunswick

We also wish to warmly and gratefully acknowledge many other people who assisted and encouraged us in this endeavour. A special thank you to Julia Cochrane, copy editor, who read the manuscript with an eye for accuracy. Paul Fam, our publisher, has brought enthusiasm and an unwavering emphasis on student learning as the fundamental purpose of our collective endeavour. Candace Morrison, our developmental editor, shared her fascination with our multiple disciplines, which has given her a particular interest in our book (and engaged her science background) and a singular purpose: to keep us focused on the matters at hand and the urgencies of the schedule. Thank you for your unfailing courtesy, even in the midst of a firestorm! We're also grateful that you share in our curiosity about science. Your efforts have been a major factor in the coming to fruition of our writing project. We also applaud the unsung but absolutely indispensable contributions by those whose efforts transformed a rough manuscript into this final product: Dianna Little, cover and interior designer; Sean Chamberland and Leanne Newell, marketing managers; Melody Tolson, permission and photo researcher; and Erin Moore, proofreader. We'd especially like to thank Benjamin Kelly. Included is a special thank-you to James Gage, who thought up the title for our book.

Fiona Rawle thanks her husband, Adrian, and their three children, Lucas, Anna, and Emmie, for their support, patience, and constant smiles during the completion of this book. Fiona would also like to thank her students, who are a constant source of inspiration to her.

Todd Nickle thanks his wife, Penny, and their two children, Tim and Erin, for their patience and encouragement as he eked out moments to write and proofread these pages. He also wants to thank his students for their enthusiasm and persistence even as he increasingly challenges them throughout their semesters and academic careers.

Rob Thacker thanks his wife, Linda, for her patience and almost limitless support. He also thanks his colleagues Dr. Adam Sarty and Dr. Roby Austin for stimulating discussions about education techniques, as well as the students of the SMU Astronomy and Physics program for providing some incredibly helpful editorial input.

Susan Morante is always grateful to her children, grandchildren, and students for their constant delight in the world that surrounds us, which provides the inspiration for all that she does.

DEDICATION

To my current and future students. And to "Grumps," my first teacher. ~ Fiona

To my family, Tim, Erin, and Penny, and to all those students who also bring joy to my life, making work and play sometimes hard to distinguish from each other. ~ Todd

To Dr. George Jaroskiewicz, for being an inspiration and for teaching me that physics and the "real world" are one and the same. ~ Rob

To my family and past and future students, with the desire that the curiosity and fun of science never be lost. ~ Susan

1

Purpose of *Science*[3]

What are the things you find most beautiful in science?

"Science is beautiful when it makes simple explanations of phenomena or connections between different observations. Examples include the double helix in biology, and the fundamental equations of physics."

STEPHEN HAWKING, *The Guardian* (15 May 2011)

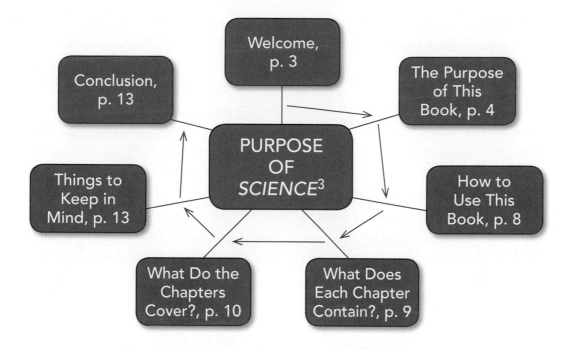

1.1 WELCOME

Welcome! The fact that you are reading this book right now means that you are either in the midst of taking **science** courses or preparing to take them in the future, you want to understand science, and you want to do well in those courses. The best time to be reading this book is NOW. The sooner you develop effective scientific thinking, studying, and communication skills, the better off you'll be in your university studies and your career.

The Authors

This book was written by four university professors who teach undergraduates in the physics, chemistry, and biology disciplines of science. As educators, we care about how students learn about science and are actively involved in figuring out better ways to teach and learn scientific material. This means that we are interested in **pedagogy**: the science of education. We often hear from our first-year science students that they feel overwhelmed by the amount of material they need to learn, under immense time pressure, within just a few short months. We also get feedback from students who feel as if learning science is like learning a new language, with massive lists of scientific terminology, and learning to read graphs is like interpreting hieroglyphics. This book is our response to these concerns. There is a better way to study science—resulting in true **understanding** rather than rote memorization. This method is based on learning how scientists think, using the language of science and focusing on the connections between scientific disciplines. We wrote this book to help you learn science in a way that makes sense and will help you excel in science courses both now and in the future. We view this book as a conversation between us, the professors, and you, the student. This book is intended as a personal conversation between a student who is interested in getting advice and an educator who wants to give the best help possible to that student. To better achieve this feel, instead of using "we" (as in the four professors, which sounds imperious and distant at best), the book is written in the first person, such as "I want you to do really well in the sciences," rather than "We want you to do well …."

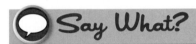

PEDAGOGY: The word *pedagogy* originally comes from the Greek word *paidagogos*, which means "teacher"; this word in turn comes from the words *pais* ("child") and *agogos* ("leading"). Pedagogy is the science of how people learn.

1.2 THE PURPOSE OF THIS BOOK

Why are you reading this book? Why are you holding it in your hands right now? Why should you care about improving your science study skills? Perhaps you are getting ready to take science courses in a few months, perhaps you are in your first term of university or college, or perhaps you're farther along in your academic career. Whatever your motivation, this book will help you to understand your science courses and study effectively. The purpose of this book is to bring the disciplines of physics, chemistry, and biology together, teaching you how to understand the sciences as a whole (rather than as discrete units) and how to study these subjects in a way that will enhance your understanding, not only for your current course, but also for all your remaining courses at university.

The Main Goals of *Science*[3]

The main goals of *Science*[3] are as follows:

1. To provide you with a quick and easy-to-understand "hub" of material that will help you succeed at university.

2. To bring together what professors want their students to know and what students feel they need to succeed.

3. To teach you how to think like a scientist.

4. To give you the skills you need to succeed at university, focusing specifically on study skills, writing skills, content-processing skills, and numeracy skills.

5. To target common student weaknesses using proven strategies and methods.

6. To integrate the major branches of science: physics, chemistry, and biology. It doesn't make sense to study these disciplines completely independently. They are inherently linked, and reminding yourself of these links and connections will help you to better understand them.

7. To help you to take ownership of your learning.

8. To provide a companion website that will answer any extra questions you have.

Students and professors tend to agree that the most common barriers to student success in the sciences are difficulties with critical thinking skills, time management, study skills, numeracy skills, basic writing skills, avoiding plagiarism, referencing skills, and vocabulary. This book addresses each of these categories and provides you with concrete advice on how you can improve your skills in these areas.

 Go to **science3.nelson.com** to see professor testimonials about what they wish their students knew when they entered their classrooms. You can also see student testimonials about what skills they wish they had a better grasp of at the beginning of their first year in the sciences.

No matter what techniques or strategies you decide to use from this book, you will need to put in a lot of time and work to become good at science. In fact, some professors feel that you need 10 000 hours to become good at science, which equates to roughly five years of studying science courses! There is no magic formula—no magic "download" strategy. Many students think that science is like that—a download of facts: "If I memorize this textbook of facts, then I'll be able to do science and get an A." Even if it were possible to insert textbooks of **knowledge** into your brain, it wouldn't necessarily help you "do" science. As you'll learn in Chapter 2, science is a process, not a collection of facts. The emphasis of science education is actually on skill development rather than on content regurgitation. Science relies upon creative but disciplined thinking to bring together concepts and create new understanding. The ability to do science is a formidable and highly employable skill.

Just because there is no magic solution to learning science, don't panic. Through careful analysis of professors' views, students' views, and research into how people learn, it is possible to come up with a highly effective strategy that will assist you in the process of learning science. That is the purpose of this book. You will learn that deeper learning and actual comprehension of the "bits and pieces" of science take a little more effort, but will pay dividends down the road. You will understand that working on the basic skills of math and **etymology** (which is the study of where words come from—that's in Chapter 4), as well as organizing your thinking before starting to work, will result in higher-quality understanding of the material. This translates into greater understanding, better marks, and bigger opportunities down the road in your career.

This book is not a collection of professors' opinions. Rather, it is based upon study strategies that have been proven to work. In short, it is based on *evidence*. Would you take a medicine to treat a bacterial infection that has been shown to be completely ineffective? (I'll guess that your answer is no.) So why, then, would you choose to study in a way that has been shown to be ineffective? In the same way that medical researchers do experiments with drugs to determine what works, researchers do experiments on study skills and study strategies. When a doctor treats a patient according to evidence, it is called **evidence-based medicine.** When a professor teaches a student using techniques shown to work through carefully designed experiments, it is called **evidence-based pedagogy.**

This book was modelled on the approach of evidence-based medicine. Evidence-based medicine considers not just clinical experience but also patient experience and research evidence. Similarly, this book is based on professors' experience, students' views, and pedagogical research evidence. Evidence-based medicine is the process of applying the best evidence, gained from experiments using the scientific method, to decision making in the clinical setting. In other words, it treats diseases in a way that the evidence has shown to be effective. Evidence-based pedagogy is the same thing, only applied to the way people learn. With evidence-based pedagogy, teaching and learning methods that have been shown to be effective through experiments are used in the classroom. These methods are based on *evidence*.

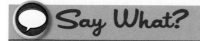

Say What?

EVIDENCE-BASED MEDICINE: Using medical interventions (such as drugs and surgeries) when evidence exists and shows that they are effective. The evidence comes from carefully designed experiments and clinical trials.

EVIDENCE-BASED PEDAGOGY: Using teaching and learning techniques that are based on evidence. The evidence comes from carefully designed and well-thought-out studies that have been conducted on how people learn.

Many years ago, information was harder to come by. Believe it or not, the concept of a "book" was considered a radical technology. (And it was! Those who could read books occupied higher social positions than the general population.) With the Internet, there's little skill in finding information. Instead the challenge is to avoid information overload and to be discriminating about what information is good and what is misleading and wrong. Remember that the Internet does not have a filter—both good and bad information will float to the top of your Google searches. Furthermore, today's scholars must also be able to integrate information from different sources and different fields. That's what this book emphasizes! Don't make the common mistake of thinking that just because you found a great website with good information your job is done. You've got to "own" the material—intellectually—not just "possess" it.

❓ *Did you know?*

Richard Feynman was an accomplished American physicist. He stated that one of the great steps in his early education was when he realized that there was a big difference between knowing the name of something, versus actually knowing something.

1.3 HOW TO USE THIS BOOK

You can read this book cover to cover, or you can jump to the specific chapter you are most interested in at any given time. If you are a "chapter jumper," however, consider reading Chapters 1 to 4 before jumping around, as this will give you a solid foundation in thinking like a scientist and understanding the language of science. Chapter 2 describes what science is; Chapter 3 teaches you how to think like a scientist; and Chapter 4 will guide you in deciphering and understanding the jargon that comes with physics, chemistry, and biology.

This book is designed not only to give you a foundation for studying in the sciences, but also to act as a go-to guide for questions you have as you take new courses. For example, if you are struggling with whether to use the word hypothermia or hyperthermia, then flip to Chapter 4, where you will find a concise list of key Greek and Latin roots used in biology. If you have to write a lab report and you don't know how to cite your sources, then flip to Chapter 8. If you have an interview coming up for a summer job and you want to prepare, then flip to Chapter 9.

In addition to being the go-to book for everyday questions that may arise in your studies, this book is accompanied by a companion website known as "The Hub." Although this book is designed to have the content you need, it does its job best if it's also **concise**. Some examples are here, but the website contains other resources. There are short videos of testimonials about the things graduating students wish they knew in their first year alongside movies that show how to make a graph using Microsoft Excel on a Macintosh or PC. Expanded writing styles, such as the CSE style (Council of Science Editors, an alternative to APA), and extended lists of concepts, should you require them, are also there for you.

1.4 WHAT DOES EACH CHAPTER CONTAIN?

As you read and explore this book, you'll notice that we have included several features to help you connect with the material. These features are as follows:

1. **Topic Maps.** A topic map is a visual/graphical means to organize thoughts about a topic. These maps give you the opportunity to look at a list of concepts in a more visual way and each chapter has a topic map to help guide you through the content.

2. **Numbered Boxes.** These boxes give simple and direct instructions on how to solve problems. For example, one problem box in Chapter 5 will be on the topic "How to remember the order of evaluations in formulas: BEDMAS." You may have learned about this in high school and will probably benefit from a clear, concise explanation of it again. These boxes give you a crash course on how to do something, with different types of examples.

3. **WWW.** This icon is to remind you that more material is available on the online website that accompanies this book.

4. **Say What?** This icon will give the definition of a word or describe the origin or root of a word. For example, *science* is from the Latin word *scientia,* meaning "knowledge" or "knowing," which ultimately comes from the Latin *scire,* meaning "to know." The purpose of this icon is to help you untangle and decipher all the jargon that accompanies scientific disciplines.

5. **Strategies That Work.** This icon highlights advice that is based on pedagogical evidence. This means that it is based on a well-designed study that was published in a peer-reviewed scholarly journal.

6. **Did You Know?** This icon highlights cool science facts, linking book content to other interesting areas.

7. **Danger Zone.** This icon highlights things that students do that sometimes get them in trouble, or study strategies that students often use that have been shown not to work.

8. **Tool Kit.** This icon outlines specific tools (including links and web resources) that you may find useful. As you explore this book, you may want to visualize a toolbox that you will be filling up with skills and techniques that will help you to study science.

1.5 WHAT DO THE CHAPTERS COVER?

The topics that appear throughout the book have been chosen based on professor and student discussions, published studies on pedagogy, and research into undergraduate science curricula at universities and colleges in North America.

Chapter 2 What Is Science?

In **Chapter 2, What Is Science?** you will learn that science is a way of knowing about, and exploring, the world. You will see in this chapter that science is a process, rather than a collection of facts that you need to memorize to pass your next exam. This chapter lays out the key elements that link different disciplines of science, such as physics, chemistry, and biology, together. When you finish this chapter, you will see that your science courses are creatures of interconnected ideas.

Chapter 3 Thinking Like a Scientist

In **Chapter 3, Thinking like a Scientist,** you will learn how scientists are driven by curiosity. They are often intrigued by the world around them, which they regard with the eyes of novices. Science often relies upon hypothesis testing. Scientists create a hypothetical explanation—a hypothesis—that can then be tested to see if it can be maintained, modified, or even discarded. Hypotheses are valuable even if they're disproved, as they can be refined to allow different ways of thinking to be tested. When you finish this chapter, you will understand that scientists use science as a tool to find out more about the natural world. Ideas are often subjected to tests called "experiments." You will also learn about the fundamentals of critical thinking and the components of a good question. You will see that science is about asking questions, and you will learn how to ask the right questions.

Chapter 4 Understanding Science

Chapter 4, Understanding Science, is a key chapter of this book, and you should expect to access it frequently. Its purpose is to enhance your understanding of your course content. Much emphasis is placed on language and terminology—by describing and profiling keyword roots, the chapter will help you to easily understand course language. When you finish this chapter, you will understand the basis of scientific terminology and be familiar with keyword roots. You will understand where the language of physics, chemistry, and biology comes from. Being conversant with proper scientific expression will make your job easier as you go through university—and far beyond!

Chapter 5 Science and Math

Chapter 5, Science and Math, goes into detail about the math skills you need to excel at science. Many students are nervous about mathematics, and this chapter will help to calm those nerves. At the end of this chapter, you will have learned that math, like terminology, is a tool. It can add clarity to our understanding of natural processes. It helps us solve problems more quickly and accurately. In this chapter, you will develop your skills in identifying units, solving fractions, and performing simple statistics with online quizzes that present information in a variety of ways, including algorithmically generated variations.

Chapter 6 Physics, Chemistry, and Biology Basics

Chapter 6, Physics, Chemistry, and Biology Basics, is a brief overview of key biological, chemical, and physical concepts, connecting these three disciplines. It starts with a discussion of the connections between the sciences and looks at the relationship between course content and everyday life. It then progresses to identify the top concepts for physics, chemistry, and biology courses. Discussions with professors and a detailed review of first-year science curricula at North American universities were undertaken to come up with the lists in this chapter. At the end of this chapter, you will see that each discipline consists of more than just a list of facts, and that some key concepts across different science disciplines are interconnected. Nature isn't itself organized into chemistry, biology, and physics, but instead it is studied by scientists who investigate smaller features within what we broadly break up into these specialties. Physics informs chemistry; chemistry informs biology, for example.

Chapter 7 Study Strategies

In **Chapter 7, Study Strategies,** you will be guided through a description of study techniques that have been shown to be effective in the university setting. Unlike the techniques that probably worked well in high school, study strategies in college and university require much more synthesis on your part. Memorizing lists of information is often a path to poor performance at higher academic levels. Rather, it's important to be able to understand and explain how different concepts connect. Being comfortable with the language of science is important. Recognizing word roots and context clues will greatly enhance your ability to prepare appropriately for class. At the end of this chapter, you will realize that you must invest time and effort out of class in order to achieve the best possible **learning outcome**. Simply reading the text, making flashcards, and memorizing lists won't be sufficient for outstanding performance. Instead, you need to be reflective about techniques and behaviours that you will employ in order to understand the material. Understanding how knowledge is generated and expanded is a critical tool for your success.

Chapter 8 Science and Communication

In **Chapter 8, Science and Communication,** you will be led through a quick and easy-to-understand guide to writing, focusing on the basic skills for improvement. This chapter contains a description of summarizing versus paraphrasing versus plagiarism, discussing why plagiarism is unacceptable. This chapter also describes how to write a lab report and how to reference sources properly. It is designed to provide you with not just a guide to better writing for one course, but a reference for your whole academic career. At the end of this chapter, you will know how to approach scientific writing assignments by breaking the task into small discrete steps, and how to communicate scientific knowledge appropriately without plagiarizing.

Chapter 9 Job, Scholarship, and Post-Graduate Applications

In **Chapter 9, Job, Scholarship, and Post-Graduate Applications,** you will get a crash course on CV preparation; scholarship, job, and post-grad applications; and interview preparation. This chapter is included because student surveys reveal high demand for clear instruction on these topics. Keep in mind that the main reason the end of this book is about post-graduate and job applications is that you will eventually have to apply what you have learned. Thus, throughout your university career, you want to focus on the learning, rather than just the grade you get (more about that in Chapters 7 and 9).

Chapter 10 Conclusion

Chapter 10 is the **Conclusion** of the book. It highlights the big picture and sums everything up. It is followed by the **Appendices,** which contain an overview of scientific literature and a summary of key tables from the book. The appendices are an easy-to-go-to section, containing information that you will need rapid access to whether you are in class or studying at home.

1.6 THINGS TO KEEP IN MIND

Keep the following in mind as you read this book. As you'll learn in Chapter 2, science is a process of learning about the natural world. It is not a list of facts to memorize. So think of "learning science" as "developing skills for connecting ideas and relating facts," rather than just memorizing lists. This book will help you to develop skills associated with success in studying. Many of these skills are considered to be **foundational skills** (also called **transferable skills**)—meaning that these skills will be built upon throughout your university or college career. You will need them later in life. There is life beyond university or college, and that is where you'll need to apply what you've learned to be successful.

1.7 CONCLUSION

Let's begin! Chapter 2 will set the stage as we look at what science "is"!

What Is Science?

"Research is to see what everybody else has seen, and to think what nobody else has thought."

ALBERT SZENT-GYÖRGI

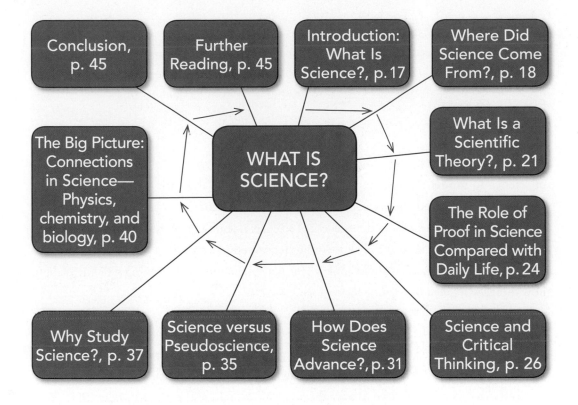

2.1 INTRODUCTION: WHAT IS SCIENCE?

Let's start with what science *isn't*. It isn't a long list of facts you have to memorize. It isn't something that never changes or can't be challenged. It isn't something that only people in lab coats know. And it definitely isn't stored in books or in TV shows.

Science *is* a way of discovering the nature of the Universe around you. It combines a shared body of knowledge with a method of investigation to create one of the most powerful tools known to humanity. From collected data, scientists are able to infer the behaviour of systems using logic and mathematics. Whether studying atoms, amoeba, or asteroids, the same overall approaches are applied even if the equipment or questions being addressed are different.

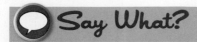

Science also makes testable predictions. Day in, day out, our society relies on the predictions of science to **model** things from the spread of influenza to the paths of storms. But, most importantly, these predictions can also be used to test the accuracy of science. This property is perhaps the key contributor to the success of science. If you can test the accuracy of scientific predictions, then you immediately know whether something is worth using or not. In fact, the edge of scientific research is about looking for new phenomena, or forming new ideas about how things work, *and then testing them.*

The amazing success of science is all around you, and it's virtually impossible to avoid. Just think about getting up in the morning. The pattern on your comforter was created with artificial dyes whose discovery dates back to the 1850s. Your toothbrush is made of plastic and synthetic fibres, both of which can trace their development to the beginning of the 1900s. Your toothpaste has undergone a barrage of scientific testing and contains chemical salts designed to prevent the growth of bacteria (so your breath is fresh!). If you have a bowl of oatmeal, you will likely use a microwave oven. The discovery that microwaves can heat food happened by complete accident when Percy Spencer, a radar technician, discovered that his chocolate bar started to melt when it was close to some equipment he was building.

These simple examples only just touch on the impact of science. If I include the discoveries that have made detailed engineering and technologies possible, then it becomes clear that science has truly shaped the modern world. But the steps that lead up to the science of today can be traced back thousands of years.

2.2 WHERE DID SCIENCE COME FROM?

A precise definition of science is pretty difficult—in fact philosophers of science have argued about it for years! For the moment, let's expand the idea of science to include investigation of the world around us. That has a very long history. In fact, humanity's knowledge of how and when this kind of investigation started is really limited by when written records and language became common. For events that occurred before reliable records were kept, we have to rely on some pretty good guesswork to put the pieces together. But it is very clear that a number of Neolithic monuments, such as Newgrange in Ireland, which was built around 3200 BCE, are aligned precisely to match seasonal events.

The earliest form of writing appears to be Sumerian cuneiform around 2900 BCE, and shortly after that Egyptian hieroglyphs appeared. Writings from these times show that the Sumerians, Babylonians, and Egyptians all had advanced knowledge of astronomy and could predict a number of celestial events. Just like the Neolithic people in Europe, their immense ceremonial buildings, ziggurats, and pyramids were usually aligned with specific celestial events or objects. And if you ever wondered where the units of 60 for minutes and seconds came from, you can thank the Babylonians! With this capacity for analyzing and predicting the world around them, it's no surprise that intensive agriculture and other technological developments like the wheel can be traced back to these societies.

Perhaps the most famous pre-science investigators were the ancient Greeks. The Greeks prized thought and philosophy, although it is quite ironic that one of their most important thinkers, Socrates, is said to have believed that writing was actually a bad thing. Why? Because he felt that it caused people to remember less and hence lack understanding. Similar arguments are made today about computers!

❓ Did you know?

Author Nicholas Carr has gathered evidence that the Internet is changing our brains in profound ways. Check out his book *The Shallows* (Carr, 2010).

Some of the most famous mathematicians can be found in Greek culture. And if you ever wondered why the Greek mathematicians were so focused on geometry, there's an interesting story behind that. As it turns out, their method of writing down numbers was so cumbersome (there were 27 different symbols!) that simple written arithmetic was often quite slow. Our modern base-10 counting system and notation is actually very efficient for adding up a large list of numbers. Hence, rather than spend their time working out numbers, the Greek mathematicians tended to focus their efforts on geometry, as the diagrams and algebra could be worked out quickly.

Other early cultures, notably India and China, also developed their own approaches to scientific inquiry. However, cultural differences played a very important role in the limits of what might be investigated, be it through religious or even philosophical considerations. Indeed, some scholars argue that certain aspects of Chinese philosophy made even the concept of a **law** of nature almost impossible for them to conceive. Despite this, the Chinese were masters of developing technologies. Wood-block printing, for example, was developed in China centuries before the printing press in Europe.

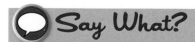 Say What?

SCIENTIFIC LAW: A description of some aspect of the world that at the time of writing has never been observed to be wrong. Laws do not have to explain why or how things happen, but must predict results. They are almost always derived from experimental results.

While Europe plunged into the Dark Ages, Islamic science grew in complexity, which some researchers argue may have given birth to the concept of the scientific method. Centuries before, the Greeks had prized thought and discussion, but Islamic scientists relied upon experiment—they had an empirical rather than a philosophical approach. The success of this method led to many, many developments in physics, astronomy, and chemistry. Unfortunately, the development of Islamic science was eventually crippled by a number of factors. While the exact cause remains a matter of debate, it seems likely that Mongol conquests, which destroyed libraries, observatories, and universities, played a large part in this decline. Just as this decline finished, however, the Renaissance began in Europe.

The Renaissance saw a return to a "classical" mode of thought. The philosophy and questioning approach of the Greeks was reignited. Universities began to spring up across Europe, and the widespread use of the printing press had an enormous impact on the spread of knowledge. Simultaneously, the **scientific method**, supported by achievements in astronomy, physics, anatomy, chemistry, and biology, showed what advances are possible through an empirical approach to investigating the Universe. This questioning approach also led to notable conflicts between scientists and religious officials. By the end of the Renaissance, the Scientific Revolution had begun. Superstition was being replaced by reasoning.

By the eighteenth century, Europe was in the Age of Enlightenment. Scientific journals began to spread new ideas more rapidly than ever before. But equally importantly, interest in science grew to new levels, becoming a topic for discussion in coffee houses. Even public lectures on science were organized and well attended by the public. The rise in popularity saw more scientists trained (the number of scientists was doubling every 15 years), and with the ever-growing interest in measuring and cataloguing, the first ideas about statistical analysis blossomed.

While the rest may be history, the growth of science in the nineteenth and twentieth centuries has had a profound shaping influence on modern society. The growth of scientific knowledge has been exponential, as has its applications—the influence of technology is pervasive. Today there are millions of scientists around the world publishing millions of scientific papers every year.

2.3 WHAT IS A SCIENTIFIC THEORY?

For all the obvious impacts of science, the vast majority of people today don't really understand how it is performed. In fact, most don't understand what words like *theory* and *evidence* mean to scientists.

A scientific **theory** is essentially a **model**, by which I mean a representation of something. For example, a toy car is a model of a real one, in the sense that the ratios of lengths (length, width, and so on) remain the same. But some aspects of a toy car aren't the same as the real one. The obvious example is that the toy car doesn't function like a real one—there isn't a miniature piston engine in it, for example.

The point of a model is that it includes features of what it is supposed to represent. While a toy car is a physical model (you can pick it up), in science we use conceptual models. Conceptual models can be mathematical, such as Newton's laws of motion, or they can be constructed in logical arguments alone, like the theory of evolution. The modelling aspect of theories is also accompanied by an explanation. For example, theories of chemical bonding are explained through properties of the outermost electrons in atoms.

This definition of a theory is distinctly different from how we often use it in day-to-day life. Although I'll talk more about this later, the most common usage of *theory* in everyday language is perhaps better described as a **hypothesis**. A hypothesis is essentially a good guess for what might have been or be happening. You might hypothesize that people eating a particular food got sick because of it. You could then test that hypothesis by analyzing the food. In general, though, a hypothesis is limited to fairly specific cases, and they aren't general enough to be very predictive.

Given this distinction, it should be clear that one of the key strengths of a scientific theory, and its underlying model, is the ability to make predictions. If your theory uses math and contains equations, you can try different inputs that represent different experiments and see if the equations predict results that compare well with the experimental results. This idea of testing theories is extremely important and is one of the central reasons why science is successful.

Scientific theories can be proved to be false. If you look at what the theory predicts and the experimental results differ, then the theory is wrong. But if the theory predicts something and that's what you observe, does that mean the theory is correct?

At first it's natural to think "of course!" But things are more complex than that. Whenever you observe something—even if you're being diligent and making careful **measurements**—there will be **uncertainties**, which are sometimes called "errors." For example, you can't measure anything perfectly accurately. Consider measuring the length of a pen with a centimetre ruler. You can only estimate the location between the centimetre markers at which the pen's length ends. It doesn't matter how fine you make the scale on the ruler, you'll always be unsure where the exact length is between the scale markers. This error might be incredibly tiny if you have a really fine scale, but it will always be there.

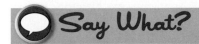

Say What?

UNCERTAINTY: The amount of error in any given measurement. While it may seem annoying that you need to calculate it, the fact is that it is one of the most important tools in the scientific method.

These measurement errors have a big implication: You can never be absolutely certain that a scientific theory has predicted the right answer. But you can know when it has predicted the wrong answer.

At first that doesn't sound like science at all. Doesn't science tell us when theories are right or wrong? Doesn't science *prove* things to be right?

Stop for a second to think about how you learned science in high school. When you first met gravity, it was described as "uniform acceleration of 9.81 m s^{-2} near the surface of Earth." By the time you left high school, you may have been taught that this isn't universally true— gravitational acceleration changes depending on how far you are from a given object. So you've replaced one theory that works pretty well in some cases with another that works for many more, and is more accurate.

But even Newton's theory of gravity gets some things wrong. Einstein figured this out about a hundred years ago when he developed his theory of general relativity. If you want to describe the effects of gravity near a black hole, for example, you'd better be using Einstein's theory.

You might well be asking: "Have we been able to show Einstein's theory is wrong?" Not yet, but it's the subject of a lot of research!

I hope by now you have a picture that science knowledge builds up through a process of replacing less accurate theories with more accurate ones. But just because you have a new theory, you don't always get rid of the old one because it gets a few things wrong. It may be simpler to use, or it may give answers that are "good enough." As an example, Newton's theory of gravity is more than good enough for astrophysicists to determine the orbits of satellites, and it got the Apollo astronauts to the Moon and back safely.

All of this might sound a little complicated to begin with. We don't spend most of our daily lives worrying about questions like "just how accurate is my watch?" We have a time, and we go with it. Most of the time we much prefer having true/false–type questions, or answers that we think are absolutely right. Is it real or is it fake? The idea of something being right, or being able to "prove something," is an idea we all think we understand—or do we?

2.4 THE ROLE OF PROOF IN SCIENCE COMPARED WITH DAILY LIFE

This is one area of science that is frequently misunderstood. The problem is that the concepts of proof and theory are used in daily life in dramatically different ways to their scientific meanings. So if you want to get a better idea of what science is, it really helps to contrast it with other areas of our lives.

Let's consider the different meanings of proof in everyday life, starting with their use in law. Speaking as a scientist, evidence and proof in law seem particularly nebulous. We've all seen opening statements in TV courtroom dramas—a defence lawyer boldly proclaims that he or she will prove their client's innocence. But what happened to reasonable doubt? On television it is frequently out-competed by certainty, no matter how unrealistic.

In real life, forensic evidence is often riddled with doubt and uncertainty. Convictions may fail because evidence cannot be attributed to the defendant beyond a certain level of probability.

If you want unquestionable proof and complete certainty, look to mathematics and logic—you won't find it anywhere else. At its very simplest level, mathematics builds on a system of rules, or **axioms**. ($1 + 1 = 2$ is an axiom.) We don't ask *why* it is true. It is a definition—something that is unquestionably, fundamentally true. Mathematical proofs use axioms to uncover absolutely certain results. It is a comforting thought that Euclid's proof of the Pythagorean theorem will not suddenly become untrue because of some new evidence that $1 + 1$ does not actually equal 2.

Science, however, is not mathematics. You cannot prove a scientific theory is absolutely, 100 percent true. To many people outside science this comes as a surprise. Non-scientists are used to reading headlines about scientists proving the link between a cause and an effect. Yet just because you think you can see one factor repeatedly influence an event, that doesn't mean it caused it. Or, as scientists like to say, correlation does not imply causation.

This is not as damning as it sounds. Remember, I said earlier that science can prove whether theories are *false*. And that makes scientists the ultimate pragmatists. We work with theories that might be true, but we're never able to know for sure. It's all a question of statistics. Theories are accurate to a certain level of probability, or uncertainty. Knowing the uncertainty is incredibly important.

Scientists also carry some of the blame for the confusion around proof. Even though it isn't technically correct, extremely accurate theories that have not yet been proven wrong are sometimes given the status of a "law." Yet some laws are eventually shown to be wrong. Newton's law of universal gravitation is an example. Thanks to Einstein, we now have an even more accurate theory of gravity, general relativity. But we keep on saying "Newton's law" because it very accurately describes gravity in many situations.

So science is really a very down-to-earth pursuit. It's a logical and pragmatic approach to examining the world around us. The key point is weighing up evidence and putting the facts together in a way that doesn't reflect your opinions or thoughts about the way things should be. And that can be difficult to do!

2.5 SCIENCE AND CRITICAL THINKING

Are there any ways in which we naturally "think scientifically"? If, for a second, we generalize thinking scientifically to mean systematically asking questions and looking for answers, then yes, there are number of examples of how we do that in everyday life.

How did you choose your courses at university? As long as you have some flexibility in your program, you likely took into account a number of factors. Does this class give me good foundational skills for ones that follow? How much do I enjoy the subject? Does the class fit in with the rest of my schedule?

You probably go through a similar process of asking questions, a *critical analysis,* when you consider buying something expensive. That is, unless you're someone who likes to spend their money without really thinking about whether you're getting good value! Because money controls a lot of what we can do in our lives, it's no surprise that people carefully consider many purchases, often with the help of one of the thousands of websites dedicated to providing advice on buying one thing or another. The same thing goes for choosing a university or career options.

The step-by-step approach of looking at options and eliminating things that do or don't work is really just one part of the much larger framework of **critical thinking**. Critical thinking relies heavily on **logic**, **honesty**, and **falsifiability**. To help people understand the process, James Lett developed a simple acronym that describes the six key parts of the critical thinking process: **FiLCHeRS** (Lett, 1990). The letters in the acronym come from the following key points.

FiLCHeRS

FALSIFIABILITY. I've already presented the importance of being able to deduce that something is not true. If there is no experiment or evidence that would falsify a claim, then essentially you have to treat the claim as meaningless. This is actually a very powerful idea. If you think about the claims of people who say they have seen pixies, they will argue you can only see them on certain days, or only certain people can see them, or you can't even take pictures of them. The idea itself is being framed in a way that makes it impossible to disprove. Also, some claims can be so vague that no possible link to measurable things can be made.

LOGIC. Any argument must be explained in terms that can be rationally followed from the underlying premises. If any step in an argument cannot be justified, then the whole argument is invalidated. Some failures of argument are obvious: for example, the argument "all cats eat mice; Sarah eats mice; therefore Sarah is a cat," fails because although all cats eat mice, other animals do as well. Many times, however, the failure of the argument is much more difficult to determine and frequently takes a lot of time to uncover.

COMPREHENSIVENESS. Most scientific theories actually make different kinds of predictions depending on the circumstances. Sometimes scientists find some results that support the theory and others that don't. In this case, the unsupportive results must be given the same importance as the supportive ones. The problem can actually be very significant when marketing products. Some marketers will attempt to sell an item based on a particular result (for example, a stain cleaner is shown to remove a beet stain, suggesting it should be good on all stains), while in reality the product does not work as generally as suggested (people who buy the cleaner discover it *only* works well on beet stains).

HONESTY. Any system of critical thinking must examine evidence in an honest way. But what do I mean precisely by "honest"? In essence, the evidence must consider in a way that does not impose any expectations of what the result should be; that is, the data should be considered in an unbiased way. If the results of an experiment do not support your idea, then your idea is false. Things are in fact a little more complicated than that: You'd probably try the experiment again to check, but if the results, within the bounds of experimental error, don't support your idea, then your idea is wrong.

REPEATABILITY. One of the key problems with performing only one experiment is the possibility of error. Perhaps a piece of equipment was not set up correctly, or perhaps there was a power spike that made a reading higher than it should be. Many things can go wrong in an experiment. Checking results repeatedly is required to ensure that the same answers are reproduced, as well as being a very important part of ensuring that the answers we get aren't due to some kind of coincidence. This is one of the key points of science: A scientific theory needs to be applicable from day to day and from place to place. If the rules of science changed every day or were different in different places, we would not be able to use science to make predictions. Science would then be a bit like stamp collecting—it would be possible to catalogue results and events, but there would be no way to connect results through theories. Robbed of the ability to make useful predictions from theories, science would become impotent.

SUFFICIENCY. The data or evidence used to support an idea or theory must provide adequate support for the idea/theory. This is a bit tricky, but it brings together three important concepts: Support for an idea must be provided by the person that claims a particular result; the more important a claim/result is, the more exacting the evidence must be; and the evidence provided must be amenable to testing and should not be based solely on testimony. The first requirement reflects the fact that just because you cannot show something to be false does not necessarily mean it is true. The second requirement can be understood as, "Extraordinary claims require extraordinary evidence," while the last one reflects the fact that human beings are fallible, and, unfortunately, they might lie or be deceived.

Taken together, FiLCHeRS provides anyone with what Carl Sagan called a "baloney detection kit" in his book *The Demon-Haunted World* (Sagan, 1995). Faced with someone making a particular claim, you can apply each of the steps to see how it stands up.

To show how useful this is, let's borrow an example from the *British Columbia Medical Journal*. In an article written in 2007, Dr. Shalinder Arneja outlined how the chiropractic view that the body can naturally immunize itself stands up against traditional approaches to immunology (Arneja, 2007). It's worth emphasizing that not all practitioners of chiropractic treatment have the same view on this issue. So you should carefully separate this discussion of the theory from the actions and views of individual chiropractors that you may meet.

A FiLCHeRS Example

The originator of chiropractic treatment was Daniel David Palmer. While there is a vast body of literature associated with the philosophy of chiropractic treatment, I will focus on one specific aspect: that all living things have a spiritual healing force that can self-heal. Palmer's theory of treatment is that by improving nerve conduction, this "innate intelligence," as he called it, can cure ailments. The impediments to nerve conduction are provided in this theory by "vertebral subluxations," which are argued to be functional or structural compromises that affect the operation of the nervous system. The argument against immunology is then that by applying chiropractic approaches to improve the body's supposed self-healing ability, immunology should not be necessary. Let's examine how the concepts of self-healing, and its impact on the choice of medical treatment such as immunization, stand up to an analysis within the FiLCHeRs framework.

FALSIFIABILITY. How can we falsify the idea that living beings have an innate intelligence that can cure diseases? This is surprisingly difficult. If the individual does not recover, it can be argued that the wrong chiropractic approach was applied. Whereas if the individual was given immunization and recovered, then it can be argued that the innate healing was somehow triggered. Even a study comparing two groups cannot solve this issue.

LOGIC. The fundamental basis for this chiropractic idea is that individuals with subluxations in their spines are less immune because their self-healing power is affected by poor nerve function. If an individual with a strong immune system can be shown to have a subluxation, then the entire premise is falsified. This issue is also made more difficult by the lack of a definitive diagnostic agreement on what a vertebral subluxation actually is.

COMPREHENSIVENESS. Evidence supporting the chiropractic position on immunization is largely hearsay and anecdotal. This belief does not come close to the vast amounts of medical literature showing the direct connection between immunization and disease resistance.

HONESTY. Despite widespread evidence that immunization is a highly effective public health method, the International Chiropractic Association advocates that it should be an issue of individual choice (and thus disputes the effectiveness of immunization). Note that the Canadian Chiropractic Association *does* recognize the health benefits of immunization (which do on occasion have noted side effects, although the overall public health benefits outweigh these small risks).

REPEATABILITY. The review of medical literature conducted by Arneja found no papers undertaking clinical studies on the chiropractic position versus the standard immunization approach.

SUFFICIENCY. Arneja found only 17 articles related to immunization and chiropractic medicine. All those articles were testimonials or surveys. In the absence of evidence of a similar nature to that associated with immunization, it is again difficult to accept the chiropractic position on immunization.

From this evaluation it is easy to conclude that the chiropractic philosophy toward immunization is difficult (at best) to support. Without strong evidence in support of this position, it should be rejected. But this is not to say that there aren't other aspects of chiropractic treatment that are known to be of significant benefit, such as relieving lower back pain and easing muscle spasms.

Nonetheless, this is a powerful example of how the FiLCHeRs methodology can be applied. While not every situation allows all the aspects of the FiLCHeRs approach to be applied, using critical analysis to consider general situations or circumstances can be very helpful. If you cannot be rigorous, the key point in a critical analysis is that you ask questions about something and try to get answers. Is something logical? Does a result seem to hinge too much on someone's opinion? And sometimes, to ensure that you are honest, you even have to question what you think you know.

The process of questioning what you know can actually be quite challenging. We all like to think we know the answers. But sometimes we meet new situations where what we thought we knew doesn't apply any more. For example, having grown up on Earth, we all know that if you throw something upwards, then gravity will bring the object back down. But if you do this in deep space, far away from the gravitational field of other bodies, then what you throw "up" will just keep on going. Faced with that scenario for the first time, you'd have to conclude that "What goes up must come down" isn't a very good **scientific law** at all. It entirely depends on the gravitational field you are in, and how you are moving within it. Newton's theory of gravity is a much better scientific theory, but that too is known to break down in certain circumstances.

So science proceeds through a series of improvements. But how do these improvements come about? Are they sudden changes that shake the foundation of what scientists understand, or are they subtle improvements that take a while to take hold?

2.6 HOW DOES SCIENCE ADVANCE?

Despite a perception that science moves forward with sudden discoveries that overturn everything before them, in fact the process is usually much more gradual. A lot of scientific work is focused on understanding current theories in more depth and extending knowledge through a series of distinct but small improvements. Of course, that isn't to say that big shifts in thinking don't occur. They do! But the sheer difficulty of scientific research means that revelatory theories such as Darwin's theory of evolution and Einstein's theory of relativity only come along rarely.

Scientific advances are published in specialized scientific journals. University libraries are full of specialized journals for specific subject areas, while more general journals, such as *Nature* and *Science,* can be found on the shelves of your local newsagent. What is perhaps less well appreciated is that every article (more correctly "paper") published in these journals has undergone **peer review**.

> **?** *Did you know?*
>
> The Thomson Web of Science database tracks over 600 journals devoted to covering research in physics, chemistry, and biology.

Peer review is a way of trying to stop wrong answers from being published. Without it, people can make grand claims that could be completely wrong. Perhaps the most famous example of this is the "cold fusion" saga of 1989 (see Box 2.1 for an explanation of what cold fusion is). Two scientists released news directly to the press that they had found a new method for potentially generating limitless power. As things turned out, their results were completely bogus.

The cold fusion episode happened because the press picked up on an article that simply had not been checked for correctness and validity. Events like these, in a nutshell, are what peer review tries to prevent.

BOX 2.1   What Is "Cold Fusion"?

Nuclear fusion is a form of nuclear power in which nuclei of atoms are forced together, causing energy to be released. It's the same mechanism that powers the incredibly hot cores of stars, including our Sun. Reproducing this in simple laboratory equipment, much more cheaply than in large, inefficient, and very hot fusion reactors, could potentially revolutionize world power production. The term *cold fusion* is given to equipment that produces fusion power at a much, much lower temperature than the tens of millions of Kelvin in the core of a star or fusion reactor.

The peer review process works in the following way: The editor of a scientific journal will take each paper he or she receives and send it to a (usually) anonymous **referee**. The referee is meant to be a scientist who is respected and known to be an expert in the particular subject area of the paper. Of course the referee will not know the answer to the scientific question that the paper is trying to address—unless it has already been discovered (believe it or not this does happen occasionally).

Referees have to be chosen very carefully. They are supposed to be unbiased. That means they can't be friends with anyone who wrote the paper, or for that matter have any personal conflicts with them. If they were friends with the authors, they might let some errors slip past them, or, alternatively, if they didn't like the authors they might be especially difficult about one point or another. The referee should also not be seen to benefit from the publication of a paper. Suppose a paper is written supporting Professor Kringle's theory; then any paper that also supports it in a different context shouldn't be reviewed by Kringle. Clearly, referee selection must be done very carefully.

? Did you know?

If you have co-authored a paper with someone, then you can't referee his or her papers in the future.

Once the referees receive a paper, their job is to check the results in the paper as best they can. For papers that use a great deal of math, for example, that may mean working through the equations in the paper and checking that they are correct. In laboratory-based science the review often involves checking that the experimental methods are appropriate and accurate. The paper must also reference earlier work appropriately.

This part of the review is done as well as the referee can. Thus, there is no guarantee that everything is correct in published papers. Referees usually can't redo experiments, for example, and mistakes do get made during the refereeing process. But provided the paper reaches the appropriate standards, it is viewed as passing this part of the review process. The referee must then decide on what is called *relevance*.

The question of relevance is perhaps the trickiest part of the scientific reviewing process. To make this decision, the referee must ask, "How new or original is the work? Does it make a significant enough step forward? Does the work fit the subject area of the journal that it has been submitted to?" The latter is usually easy to answer; the two former queries much less so.

The key problem with relevance is that it is *subjective*. It requires the referee to form an opinion rather than looking at whether something is accurate or not. As you read this book, remember that science requires that data and experimental results be treated in an *objective* fashion. And here is where there can be a few problems with the reviewing process. Some people will view a small step forward as being very significant, particularly if they work in that particular area, while others may think it is not important.

Criticism of the peer review process usually focuses on this aspect. Scientists who cannot get a particular paper published have been known to say that unscrupulous referees are using this aspect of the refereeing process to block publication. The good news is that scientists can ask for a second referee, and some journals do this as a matter of course, or they can try a different journal.

The other aspect of peer review that receives criticism is the fact that referees make mistakes. It is virtually impossible for referees to check every part of a paper, and incorrect papers do occasionally get published. Unfortunately, there seems to be little that can be done about this. Research is by its very nature difficult. Also, every now and again, important results may be rejected as well. Usually, though, important work is compelling enough to get published in the end.

☠ DANGER ZONE

Just because scientists are human and sometimes make mistakes doesn't mean that all the science they create is somehow flawed. That's a logically flawed statement called an *argument ad hominem,* where the failings of an individual are used in an attempt to negate other truths.

The key aspect of peer review is that it acts as a good first pass on the quality of research. Trying to filter through all the papers submitted to journals would be a challenging task. It is, however, true that the value of papers is measured largely by how many people cite them in later works. In general, a great many papers are published that attract comparatively little attention. That's a reflection of reality: There are a great number of books that rarely get read, TV shows that rarely get watched, and artwork that goes unappreciated.

I've spent a long time discussing scientific advances and peer review because it's so important in the scientific method. Science is, by and large, something that moves forward in carefully planned steps. But luck, or serendipity, has played a big role in the history of science. In Box 2.2 I list my favourite accidental discoveries.

BOX 2.2 Top 10 Accidental Discoveries in Science

10. Dinosaur extinction asteroid crater (Penfield and Hildebrand)
9. Helium (Janssen and Lockyer)
8. X-rays (Röntgen)
7. Synthetic dyes (Perkins)
6. Cosmic microwave background (Penzias and Wilson)
5. Viagra
4. Vulcanized rubber (Goodyear)
3. Microorganisms (Van Leeuwenhoek)
2. Vaccination (Pasteur)
1. Penicillin (Fleming)

I hope by now you've got a good idea of what science is and how scientific knowledge grows. But before I go into why it's good to study science, let's look at approaches to explaining the Universe that are less than scientific, so-called **pseudoscience**.

2.7 SCIENCE VERSUS PSEUDOSCIENCE

In a world where an increasing number of TV shows focus on pseudoscientific concepts such as astrology, extrasensory perception, and ghosts, how can we distinguish phenomena that have been scientifically analyzed from those that haven't?

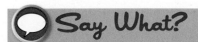 Say What?

PSEUDOSCIENCE: From the Greek *pseudes*, meaning "false," a term given to ideas or phenomena presented in a way that is not consistent with the scientific method, while attempting to appear so.

The concept of pseudoscience is comparatively new. It is used to classify an idea, or phenomenon, presented in a way that is not consistent with the scientific method. However, before the scientific method was developed, science and pseudoscience often went hand in hand. Isaac Newton, famous for many scientific advances (including the first mathematical description of gravity), also actively practised alchemy: He spent years searching for ways to turn lead into gold!

The truth is, even today, it can sometimes be far from easy to tell whether something is scientific or not.

Pseudoscience uses a number of different approaches to present a veneer of reliability. These techniques can range from mimicking a scientific approach to relying on an authority to provide confirmation or evidence on a given issue. Even for experts in science, there can, on occasion, be times when it is difficult to tell whether something has been reliably tested or not. While there is no perfect toolkit for determining whether something is scientific or not, a number of things may raise warning signs.

As I discussed earlier, the key factor in scientific advancement is the analysis of results. After accounting for possible errors, the interpretation of results must be performed without bias or preconceived notions. This can be one of the hardest parts of science, since we are all human and very often want something to be true or false. Emotions not only make us human but also make us vulnerable to persuasion.

Pseudoscience that taps into our emotions has a strong appeal for many people. One of the most convincing arguments for some is that of the conspiracy theory. When arguments are framed in this fashion people can be easily convinced to set aside reason, because the conspiracy theorists will argue that the reasoned arguments are biased. Any attack on the theory is seen as confirmation of it. When faced with this self-protective reasoning, and individuals who provide testimony that backs it up, the emotional draw can be overwhelming.

☠ DANGER ZONE

Arguments that appeal to our emotions are often exceptionally powerful. Being scientific often means you have to cast aside emotional biases. This can be really tough!

From a scientific perspective, the key issue with conspiracy theories is that they are set up to lack falsifiability. This is clearly at odds with the scientific method. Indeed, the practice of using unfalsifiable arguments extends far beyond conspiracy theories. Pseudoscience based on unreproducible personal experience, such as alien abductions, is very common.

Another powerful technique used by pseudoscience is the use of jargon or mathematics to make something seem scientific. For people without scientific training, and even in some cases for those who have it, such arguments can appear quite convincing. A classic example of the use of jargon can be found in homeopathy. Terms such as serial dilution, succession, and potentization appear scientific, and yet the practice of homeopathy has been found, after many well-documented studies, to be no more effective at curing ailments than a placebo.

 Physics is not immune from pseudoscience either. A number of theories (mostly self-published) have used mathematical arguments about the nature of the Universe. But just because an argument can be framed in mathematics doesn't mean it's right. The key difference between pseudoscientific theories of physics and a theory that is simply wrong is the underlying assumptions. The mathematics can be perfect, but if the assumptions are outlandish it doesn't matter. But there can be a grey area: When is a theory simply wrong, versus actually being pseudoscience? In some cases, that may take time to figure out through the process of peer review in scientific journals.

 While fake articles have been published in sociological journals, there is also evidence that the same thing has happened in theoretical physics. Check out the Sokal hoax and Bogdanov affair at **science3.nelson.com**.

One last technique that some pseudoscience relies upon is the attraction of certainty. Many people are well aware of scientific studies that conclude that there is a link between two things, but the level of probability of one thing causing another is not well known. Such admissions are important statements that there is still much to learn about something, not that scientists are somehow dumb because they can't figure out the details. However, when faced with a possible link between two things versus a pseudoscience theory that claims absolute certainty of a link, many people will choose the latter. Being sure about something has a strong appeal. Most of us don't like uncertainty even though it is the nature of reality. Pseudoscience that offers certainty and absolute understanding is very attractive to many people, even if its fundamental arguments don't stand up to detailed scientific scrutiny.

Pseudoscience can be difficult to spot. But it's worth taking the effort to carefully scrutinize any claims that look too good to be true. And if someone claims they are absolutely 100 percent certain about a theory, they aren't a scientist.

2.8 WHY STUDY SCIENCE?

A surprisingly large number of people think scientific research has little bearing on everyday life. But as you know now, nothing could be further from the truth. From the packaging of your orange juice to the design of your mobile phone, science underlies almost everything you can buy. And really, if you want to have a reason for studying science, you have the most powerful one—it helps you understand the world around you. In fact, it helps you understand how the Universe works. If you ever get stranded in the wilderness with just your clothes and your wits, a bit of astronomy can help you find your way to safety. A little biology will help you to select safe food to eat. Some chemistry might help you start a fire.

Before I present some of the possibilities of careers in science, let's take some time to look at the benefits of studying science even for people with other careers.

Perhaps the most important examples of people that should have some knowledge of science are government representatives and business managers. While not every business or government decision has a scientific component, many of them do. At the moment one of the biggest issues facing society is the question of climate change. Yet, how many of the politicians involved in making the decisions understand the scientific models or possess scientific knowledge that enables them to make an informed decision with the available data?

For business managers, the situation may be different. In many technology companies, for example, the managers have risen through the ranks of the company. Indeed, in many software companies, it is common to meet managers who started by developing computer code. Nonetheless, there are still situations where managers are brought in for their business acumen, and some appreciation of the science underlying decisions and products and so on is helpful.

What about in society at large? The flipside of democracy is the power of the voter. Every one of us must take some responsibility for the government we elect (even if our decision is to not go out and vote) and the decisions it makes. So knowing something about science helps all of us play an important role in the democratic process.

Many new or improved products rely on scientific advances or breakthroughs. And it's these products, either being cheaper to produce or doing more than the previous generation, that help spur economic growth. Nowhere is this more noticeable than in the mobile phone market, where generation after generation improves performance and functionality. It is amazing to think that the computational power of advanced phones today far exceeds that of the supercomputers of the 1970s.

❓ Did you know?

The Cray-1 supercomputer from the 1970s could perform 160 million instructions per second. Third-generation iPhones achieved almost 10 times that value in 2008!

If you're wondering why most scientists choose a career in science, the answer is usually pretty simple: It's because they really enjoy it. Finding out how things work and what makes them work, as well as developing brand new ideas or pieces of technology, all have great appeal. Research on learning and education—see a great summary in Pink (2010)—has also shown that people perform much better on things they really enjoy doing. It's common sense, really. Things you enjoy are easier to remember, and you happily spend more time working on them, too.

Job prospects and salaries in the sciences are good! Table 2.1 shows the average salaries (adjusted to 2012 dollars) based on Canadian data from the 2006 Census for individuals with undergraduate degrees in physics, chemistry, and biology.

Table 2.1 Canadian Salary Data from the 2006 Census

Biology	$64 350*
Chemistry	$65 700
Physics	$81 800

*Remember that these salaries are national averages for people who have undergraduate degrees in the specific field, which means the numbers aren't for people who've just graduated. Finally, Statistics Canada data show that for all fields of study, a graduate degree increases your salary by over $10 000.

Data source: Statistics Canada, 2006 Census.

2.9 THE BIG PICTURE: CONNECTIONS IN SCIENCE—PHYSICS, CHEMISTRY, AND BIOLOGY

Traditionally, the relationship between the sciences was viewed in a hierarchical fashion: Chemistry explains biology, and then physics explains chemistry. The idea here is that chemical reactions explain the operation of living cells, while physical laws explain the interaction of atoms. Taken to its limit, this suggests that physics explains everything. This idea is called **reductionism** by philosophers.

But there's a problem with this viewpoint. Once systems become very complex, breaking their operation down into simple physics equations becomes impossible. We can't write down a single equation that explains how people work, for example. In fact, an enormous number of equations would be required. Even figuring out the precise numbers to go into those equations would be impossible! That's even before we start worrying about how to actually solve them.

This hierarchical viewpoint is not a practical way of classifying the sciences. Instead, most scientists view science as one large discipline broadly divided into physics, chemistry, and biology, although there are fields of research, such as geology, that potentially span all these areas. There is also growing overlap between physics, chemistry, and biology, as well: We frequently talk about biophysics and biochemistry, for example. So-called **interdisciplinary research** is rapidly growing in popularity as modern science develops new experimental techniques. Today the connections between physics, chemistry, and biology are arguably stronger than they have ever been. It's worth exploring some of these connections in more detail.

Biochemistry

Because many biological processes can be directly related to chemistry, biochemistry underlies much of our understanding of living things at a cellular level. Common usage of the term *biochemistry* dates back to 1903, and today it is arguably the biggest area of interdisciplinary study. From figuring out cures for cancer to understanding how some athletes can recover faster than others, biochemistry tells us an enormous amount about the processes in living things.

Much biochemistry focuses on the unique chemicals present in living systems, particularly proteins and enzymes. By speeding up chemical reaction rates, enzymes can actually make life possible, because in many cases the chemical reaction needed for life would not occur fast enough without help. These chemical reactions also contribute to processes such as brewing and wine-making. Nobel prizes were awarded in 1907 and in 1922 for discovering key steps in the fermentation process, which also happens to be one of the most important steps in the creation of energy in living cells (see Box 2.3).

BOX 2.3 Nobel Prize in Medicine 1922 to Hill and Meyerhoff: Understanding Chemical Processes in Cells

Hill and Meyerhof helped us to understand how living cells use chemical reactions to drive them. Despite this research being truly of a biochemistry nature, the prize was awarded in medicine because of its importance in human physiology.

Glucose is a sugar molecule that can be viewed as the primary energy source for cells, from which their main "fuel" is derived. The

Archibald Vivian Hill

Mary Evans Picture Library/
The Canadian Press

Otto Fritz Meyerhof

Mary Evans Picture Library/
The Canadian Press

chemical reactions by which glucose is used in cells begin with an important step in which it is converted into intermediate molecules before eventually becoming adenosine triphosphate (ATP), the molecule that allows cells to move energy between them.

Biophysics

In the traditional hierarchy of the sciences, it would be tempting to think of physics and biology as being the farthest apart. Yet, the rapidly growing field of biophysics is quickly demonstrating that the overlap of these two fields covers some immensely interesting subjects, including the fabric of life itself—cells and DNA. Thus, biophysics deals with the impact of physics within living systems and covers length scales all the way from the smallest molecules at 10^{-9} m, through to vast macromolecules such as DNA, which can be up to 1 cm long.

While the discovery of the structure of DNA (see Box 2.4) was perhaps the first major accomplishment in biophysics, today much effort in this field is focused on understanding

---

BOX 2.4   Nobel Prize in Medicine in 1962 to Watson, Crick, and Wilkins: Unravelling the Structure of DNA

DNA, short for deoxyribonucleic acid, is perhaps the single most important molecule on our planet. As the information carrying mechanism for life, it has been central to the creation and evolution of our entire ecosystem. But how did we first figure out its structure?

James Watson (b. 1928), Francis Crick (1916–2004), and Maurice Wilkins (1916–2004).

Associated Press

To determine the structure of complex molecules like DNA, we rely on a combination of careful analysis and some very educated guesswork. By the 1950s, scientists were using the physics of diffraction to study the crystallized forms of molecules. On the basis of some superb X-ray diffraction work by Dr. Rosalind Franklin at the University of London, Crick and Watson were able to build a model of DNA in its now famous double-helix structure.

One sad note about this prize is that Dr. Franklin passed away before it could be awarded, and as a result the prize was shared with Dr. Wilkins, who directed and worked in the laboratory where Dr. Franklin performed her research. Unfortunately, Nobel Prizes cannot be awarded posthumously.

---

the operation of cells. One of the most exciting fields of research at the moment relates to the operation of nerve cells and the brain.

The operation of nerve impulses was modelled in the 1950s by Hodgkin and Huxley. They conducted experiments on the giant axon nerve in squid, which at 0.5 mm in diameter are sufficiently large to enable simple experiments. After much painstaking research, they discovered that nerve impulses were propagated by movements of charge across the cell membrane. If a stimulus is applied above a certain threshold, then charge flows across the cell membrane and a stimulus is sent down the nerve synapse. This basic picture shows how physics, chemistry, and biology all contribute to our understanding.

Physical Chemistry and Chemical Physics

Broadly speaking, research in physical chemistry uses concepts from physics, such as energy, force, and thermodynamics, to study chemical systems. It is arguably the oldest interdisciplinary field, with a series of lectures on the subject first being held in 1752. Chemical physics is different from physical chemistry in that it is more focused on exploring chemical reactions through the fundamental physical forces from which chemistry is derived. The distinction between the two fields is frequently blurred! From understanding batteries to understanding the processes involved in individual chemical reactions, chemical physics and physical chemistry have helped us understand the world around us and create new technologies. The growing field of nanoscience and nanotechnology is also of greatest interest to chemical physics researchers.

Today, ultrafast laser flashes (on the femtosecond and attosecond timescale, 10^{-15} and 10^{-18} seconds, respectively) allow us to follow atoms as they participate in chemical reactions (see Box 2.5). In the future, the growing field of attosecond lasers will allow us to discern what happens to electrons themselves during chemical reactions.

Understanding the physics relevant to these processes requires the use of quantum mechanics. At the smallest scales, matter does not behave the way a football or a tennis ball does. Instead, matter can behave sometimes like a wave and sometimes like a particle, depending on how you examine it. Because quantum mechanics directly impacts atoms, researchers in chemical

BOX 2.5   Nobel Prize 1999 in Chemistry to Ahmed Zewail: Chemical Reactions on Femtosecond Timescales (10^{-15} s)

Imagine being able to control individual chemical reactions and even make molecules that wouldn't normally be formed. That's one of the long-term goals of femtochemistry.

But before steps can be taken along this ambitious path, we must first be able to monitor chemical reactions as they happen. It's worth noting that for hundreds

Ahmed Zewail

Reuters

of years, scientists thought it would be impossible to view the ultrafast motion in chemical reactions. Zewail and his collaborators showed that lasers could produce very specific and controllable electric fields so that electrons, atoms, and molecules could in fact be monitored and their interactions understood. In essence, laser technology helped develop a camera with the world's shortest shutter speed.

Over the more than 10 years since this prize was awarded, short-duration pulse technology has advanced to the point where pulses as short as attoseconds (10^{-18} s) can be made. At time scales 1/1000 that of femtoseconds, it becomes possible to see the motion of individual electrons.

physics can apply it to predict how the molecules will behave. But this advance, and the adoption of quantum mechanics in general, has relied strongly upon scientific thinking to make it happen. The behaviour of very small particles is entirely at odds with our common sense, and only careful analysis of experimental results can uncover the bizarre activities at this small scale.

2.10 CONCLUSION

Science has allowed humanity to achieve some truly incredible things. From understanding how humanity evolved, to the chemical processes that allowed this to happen, to the origin of the elements that form our basic building blocks, science gives insight like no other field or discipline. I hope this discussion of what science is has helped make clear why science is so powerful. The coupling of the scientific method and the body of scientific knowledge is the key factor. Knowledge alone is not enough. Theories need to be tested and revised if they don't work well enough.

So how to do we go about investigating the Universe around us? What do we need to think about and look out for? That's what Chapter 3 is about—how to think like a scientist.

FURTHER READING

Arneja, S. (2007). An approach to critical thinking for health care professionals. *British Columbia Medical Journal, 49*(10), 547–549.

Carr, N. (2010). *The Shallows: What the Internet Is Doing to Our Brains.* New York, NY: WW Norton.

Lett, J. (1990). A field guide to critical thinking. *The Skeptical Inquirer, 14*(4), 153–160.

Pink, D. (2010). *Drive: The Surprising Truth about What Motivates Us.* Edinburgh, UK: Canongate Books.

Sagan, C. (1995). *The Demon-Haunted World: Science as a Candle in the Dark.* New York, NY: Random House.

Thinking Like a Scientist

"You don't use science to show that you're right, you use science to become right."

RANDALL MUNROE

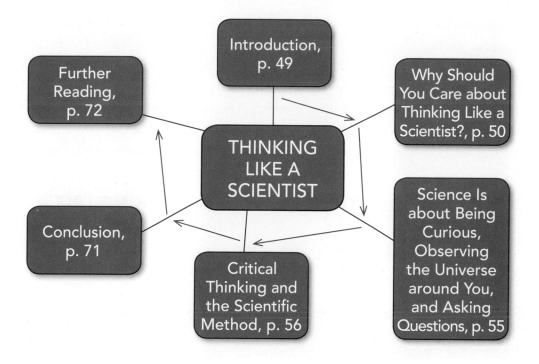

3.1 INTRODUCTION

From Chapter 2 you know that the word *science* comes from the Latin word *scientia,* which means "knowledge" or "knowing," which ultimately comes from the Latin *scire,* meaning "to know." But exactly *how* do we know? Humans have accumulated knowledge, much of it now classified as scientific, by observing the world around them and then by thinking a great deal about it. So, if you want to understand the progression of science, and if you want to be a scientist yourself (or even just achieve good grades in science classes), then you need to know how scientists think. In fact, being able to think like a scientist is useful in many different aspects of your life.

Basically, there are five things that you need to bring together in order to think like a scientist: (1) background knowledge, (2) critical thinking and reasoning skills, (3) the scientific method, (4) an understanding of science as a web of connections between knowledge and the **scientific community**, and (5) your own creativity and curiosity.

You already think like a scientist in some ways. For example, when choosing what cell phone plan to purchase, you likely compare different plans and different phones from different companies. The specifications you compare include the type of phone, battery life, the signal coverage, the cost of the plan, and whatever key features you are interested in (such as texting, voice mail, or call display). How do you make your decision about which phone/plan to purchase? You probably carefully weigh all the evidence to get the package that is right for you. You already think a bit like a scientist in some ways and already employ scientific reasoning in your everyday decision making.

3.2 WHY SHOULD YOU CARE ABOUT THINKING LIKE A SCIENTIST?

So why should you care about thinking like a scientist, and why should you read this chapter? See Box 3.1 for five good reasons!

BOX 3.1   Top Five Reasons You Should Care about Thinking Like A Scientist

1. Incorporating scientific thinking into your study strategies will help you achieve greater academic success, and using scientific reasoning skills on exams will help you achieve higher exam scores.

2. Thinking like a scientist will allow you to help solve scientific problems and be a good citizen, meaning that you'll be better able to contribute to our understanding of the world around us and participate in society.

3. Scientific thinking helps solve everyday problems.

4. Scientific thinking can prevent you from getting duped by commercial, political, and pseudoscientific claims.

5. The Universe is an amazing place, and scientific thinking will help you to understand and be awed by it.

Reason 1: Incorporating scientific thinking into your study strategies will help you achieve greater academic success.

Greater academic success means different things to different people—to you it may mean that you actually understand what you study, or it may mean achieving an A grade. Scientific thinking will help you to accomplish both. Many scientific **experiments** have been conducted, and academic papers written, on the analysis of different study strategies. Chapter 7 contains a summary of these evidence-based strategies and shows you exactly how you can incorporate them into your study plan. To be successful, you need to be critical about the study strategy you employ.

You can also apply scientific reasoning skills to your exam-writing strategy. In certain classes, it is the process of scientific thinking (and applying what you know) that is actually examined. Some students think that they can simply memorize the textbook, regurgitate it on an exam, and get an A. Low grades can sometimes be the result of bad study habits, and often students focus too much on *regurgitation* of knowledge rather than *application* of knowledge. Remember that in science, not all knowledge has been "found" yet. When training students to be scientists, professors are in fact teaching them how to discover new knowledge and think scientifically. Every answer in science leads to new questions, because there is still so much we don't know about the world. If all you have is factual knowledge, you will not excel in science. You need to be able to apply your knowledge to solve problems and discover new knowledge.

Spending countless hours memorizing scientific facts and then being able to spew them out on call does not make you an expert or a scientist. University programs are increasingly emphasizing the *process* of science and the application of scientific skills—all of which require you to *think* like a scientist (which is very different from just knowing a lot of stuff) and will better prepare you for the workforce (more on that in Chapter 9).

Reason 2: Thinking like a scientist will allow you to help solve scientific problems and be a better citizen.

If you sit down and write a list of the main problems facing global society today, the majority of the problems on your list would likely be scientific in nature. These may include climate change, resource depletion, pollution, drought, toxic contaminants in drinking water, cancer, extinction of species, the spread of disease, famine, and energy efficiency and utilization. The people who solve these problems will need to have an understanding of both science and scientific thinking. They will also need an integrative approach, meaning that biologists, chemists, physicists, geologists, and others will have to work together.

While becoming a scientist does require a strong knowledge base (in the same way as an athlete needs good core strength), it does not stop there. When you apply for a job, it is unlikely that you will be asked to recite lists of facts in the interview. You will more likely be asked to relate how you have applied problem-solving skills and critical-thinking skills, or even asked to solve unfamiliar problems (again, more on this in Chapter 9).

Reason 3: Scientific thinking helps solve everyday problems.

The utility of scientific reasoning skills goes way beyond just solving scientific problems. You can apply these critical-thinking strategies to virtually any problem, even if it seems simple or "unscientific." In fact, being able to reason scientifically about basic challenges and problems, such as what cell phone plan to buy, whether or not to buy a hybrid car, and what furniture will fit in your dorm room, will help you come to better decisions.

Reason 4: Scientific thinking can prevent you from getting duped.

Every day we are bombarded with advertisements and marketing slogans that try to convince us to buy one product or another or that make claims of "scientific breakthroughs." As a consumer, you should care about thinking like a scientist because it will reduce your likelihood of becoming a victim of some of these ads and claims. Let's look at three examples to which you can apply scientific thinking in order to avoid getting duped. For the examples below, think of how you could set up a **controlled** experiment to test the commercial claim.

THE "DETOX" FOOT BATH. Perhaps you have seen the newspaper advertisements claiming that a specialized bubbling foot bath is capable of "detoxing" your system. It works like this. You place your feet in a foot bath (which is full of clean and clear water from your tap). You plug the foot bath in and it bubbles away, allegedly detoxifying your system. At the end of the foot bath, the water has turned a dark brown colour. The advertisement implies that this dark brown colour is caused by all the toxins that have been drawn out of your body through your feet. The "before" images show clean, clear water. The "after" images show dark brown water.

Now, how could you think scientifically about this foot bath to determine if it is worth your hard-earned money? What questions could you ask? You could ask exactly what toxins are being drawn out of your body, and how those toxins are drawn out. If you kept doing the foot bath, every day for a month, would the water stop turning brown? Then ask yourself what types of experiments you could set up to determine if the foot bath is in fact drawing out toxins. What would the controls of your experiment be?

Here's one way that the claim that the brown residue contained toxins could be tested. Two baths are set up. Samples of the water are collected before plugging in each foot bath, and then samples of the water are taken during and after a detoxification treatment. However, only in one bath are feet actually soaked: The other contains water but feet are never immersed in it. The above experiment has been done, and at the end of the treatment, the water in the bath that had treated the patient's feet was indeed brown. However, the water in the "negative control" bath—the one that didn't have feet in it—was just as brown as the one that supposedly had extracted toxins in it. The water collected before the treatments looked identical, and you could—if you wanted to be rigorous—test the "before" and "after" water with a spectrophotometer or various other types of scientific equipment to gather more data. But you have one fact: The brown colour cannot simply represent "toxins."

How could this be? It all comes down to basic chemistry. The brown colour is rust (iron oxide), which forms whenever iron is exposed to air and water. Specifically, it is the iron electrodes in the foot bath that rust through a process known as electrolysis. So is this "detox" foot bath worth your hard-earned student dollars? Not unless you really are eager to make brown water and pay a higher electricity bill. But if you buy it to detoxify yourself, you'll be disappointed. It does not take a great deal of acumen to come to this conclusion—just some basic scientific reasoning.

 TOOL KIT BAD SCIENCE

The above example (and many others, including the chlorophyll example that follows) is described in detail by Ben Goldacre in his excellent book *Bad Science*. This book should be on every science student's reading list, and I strongly encourage you to take a look at it.

CHLOROPHYLL AND OXYGEN. I've seen several magazine articles, and heard nutrition "gurus," encourage the public to purchase chlorophyll pills or to eat food containing chlorophyll in order to "oxygenate your blood" or "bring oxygen into your digestive tract." What questions could you ask yourself to determine whether or not you should purchase those pricey chlorophyll supplements? Perhaps you could ask whether or not "chlorophyll provides extra oxygen for your metabolism" is good medical advice? You might even question if your metabolism needs more oxygen than you supply by simply breathing. Then you could ask whether or not chlorophyll would be able to produce oxygen in your digestive tract, or if the digestive tract can process the oxygen. (The answer in all cases is "no," as you'll see below.) To illustrate this example further, let's turn to Ben Goldacre's eloquent description (from Ben Goldacre [2010]. *Bad science: Quacks, hacks, and big pharma flacks.* London, England: Faber & Faber.):

> *"Is chlorophyll 'high in oxygen'? No. It helps make oxygen. In sunlight. And it's pretty dark in your bowels; in fact, if there's any light in there at all, something's gone badly wrong. So any chlorophyll you eat will not create oxygen, and even if it did … you still wouldn't absorb a significant amount of it through your bowel, because your bowel is adapted to absorb food, while your lungs are optimized to absorb oxygen. You do not have gills in your bowels. Neither, since we've mentioned them, do fish. And while we're talking about it, you probably don't want oxygen inside your abdomen anyway. In keyhole surgery, surgeons have to inflate your abdomen to help them see what they're doing, but they don't use oxygen, because there's methane fart gas in there too, and we don't want anyone catching fire on the inside. There is no oxygen in your bowel."*

So, oxygen + methane (+ spark) = explosion. So perhaps the very first question we should ask is whether or not you even WANT oxygen in your bowel.

SKIN CREAM AND STEM CELLS. You may also have seen advertisements for skin cream that contains stem cells to "rejuvenate and repair" your cells with "DNA technology." What questions could you ask to lead you to think scientifically about whether you should spend your hard-earned student dollars on these expensive creams? Perhaps you could ask whether the cells are alive in this concoction. If the cells aren't alive, then how do dead smushed-up cells rejuvenate your skin? If the cells are alive, how would live cells cause your skin cells to rejuvenate? And that brings us to the DNA question. What exactly is this "DNA technology," and how does it rejuvenate and repair your skin? Many of these skin creams use horse stem cells. Perhaps you could ask what the benefit of horse stem cells is over cow, pig, chicken, or rat. What experiments could you perform to determine whether or not the stem cells in these creams rejuvenate and repair your skin?

 Go to **science3.nelson.com** for detailed descriptions and flow charts of the experiments you could conduct on the above examples.

Reason 5: The Universe is an amazing place, and scientific thinking will help you to understand and be awed by it.

Exploring the natural world and the Universe has led to the discovery of some awe-inspiring and amazing things. For example, did you know that given a big enough ocean, Saturn's **average** density is so low that it would actually float? Did you know that 20 percent of the oxygen in Earth's atmosphere is made by the Amazon rainforest? Did you know that a blue whale's heart rate is about 9 beats per minute, while a human's resting heart rate is about 70 beats per minute, and a cat's resting heart rate is about 200 beats per minute? Did you know that atoms consist almost completely of empty space? The only reason things feel solid to you and me is because there are quantum-mechanical principles that stop the subatomic particles that make up atoms from occupying the same point in space! Did you know that the eight Noble metals—ruthenium, rhodium, palladium, silver, osmium, iridium, platinum, and gold—do not rust? Also, did you know that at some points on Mercury's orbit the Sun rises and sets in the same place before rising again? This is the only "double sunrise" in the solar system! There are so many interesting things like these waiting to be discovered, and it will be scientists that discover them.

3.3 SCIENCE IS ABOUT BEING CURIOUS, OBSERVING THE UNIVERSE AROUND YOU, AND ASKING QUESTIONS

Harry Brearly invented stainless steel. Albert Hofman discovered LSD. Carl-Gustaf Rossby discovered the jet stream. Melvin Calvin discovered the Calvin cycle of photosynthesis. Barbara McClintock discovered the existence of transposons. John Hetrick invented the car airbag. Jonas Salk invented the polio vaccine. Lynn Margulis came up with the endosymbiotic theory. James Elliot discovered the rings of Uranus. And Anthony Atala discovered the existence of amniotic fluid stem cells. (For interest's sake, note that the above discoveries/ achievements are in chronological order.)

What do the above scientists have in common? Curiosity! Of course, curiosity is not a trait limited to scientists. Humans are a naturally curious species. And since you are human, you are probably **curious** by nature, too. What are some questions that matter to you? What are you curious about? We've already discussed several questions that may be relevant to your life, including whether or not to invest in expensive stem cell skin creams, or what cell phone plan to purchase. What about questions that relate to your life at university or your academic success? Questions such as, "What major or honours program should I choose?" "How can I get an A in organic chemistry?" "Should I apply to graduate school?" "How can I get into medical school/graduate school?"

Every great discovery, including the ones listed above, started with an **observation**. The observations you make will lead to the formation of questions about the world around you. Being curious, making observations, and asking questions are the first steps to thinking like a scientist. They are also the first steps of the scientific method, as you learned in Chapter 2.

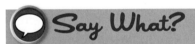

Say What?

CURIOUS: The word *curious* comes from the Latin *curiosus*, meaning "curious" or "careful," which in turn comes from the Latin *cura*, which means "care." Interestingly, the phrase "curiosity killed the cat" was first used in the early 1900s, but it first came from the phrase "care killed the cat," which was used in the late 1500s by the poet Ben Johnson.

OBSERVATION: The word *observation* comes from the Latin word *observare*, which is a verb meaning "to watch, to attend, or to guard."

3.4 CRITICAL THINKING AND THE SCIENTIFIC METHOD

In this section, we'll discuss some key attributes of critical thinking and then specifically discuss critical thinking within the context of the scientific method. In Box 3.2, I list the first three steps. Then, I explain the steps in more detail.

BOX 3.2   The First Three Steps to Critical Thinking

1. Be an active thinker and be an active participant.

2. Always ask yourself if you understand everything.

3. Question underlying assumptions.

In order to think critically, you need to:

1. BE AN ACTIVE THINKER AND ACTIVE PARTICIPANT. You need to actively have a back-and-forth conversation with whoever is trying to convince you of something, be it your professor, a salesperson, a friend, or even a scientific article. Be critical, and don't just immediately accept everything that is said. How do they know? What evidence is their opinion based on? Is it fact or is it opinion? (Recall the example of the detox foot bath, and consider the motivations of the vendor.) If you are reading a paper, you can't have a conversation with the author like you would face to face, but you can still pose questions and then see if they are answered within the paper.

2. ASK YOURSELF IF YOU UNDERSTAND EVERYTHING. Do you really? At the beginning of each introductory biology course, I challenge my students to make sure they are always asking themselves if they really understand what we are discussing (apply the who, what, when, where, why, and how questions). Meiosis is a great example. I ask my students "What is meiosis?" and I receive a textbook definition from them. Then I ask "Why does it happen?" and the students respond "to make gametes." (Note— certain life cycles are different.) Then I will ask a male and a female student, "When does meiosis happen in your lifetime? Are there differences between males and females?" At this stage several students will admit that they aren't quite sure of the answer. (Note— in women, meiosis occurs to produce eggs; it starts when the female is *in utero,* then pauses for many years, and continues just prior to ovulation.) I'm always encouraging my students to dig deeper until they are able to identify the stage at which their level of understanding changes, and you should do the same. You can find the level at which your understanding of the subject matter breaks down, and then try to increase your understanding.

3. QUESTION UNDERLYING ASSUMPTIONS. Don't take all facts at face value. Ask questions to challenge/explore scientific facts and data. Some good questions are: "Were any assumptions made?" "Was there any bias in the analysis?" "Is emotion being used in the argument?" "Are the results statistically significant?"

In Chapter 2, you learned that the scientific method is a valuable component of your tool kit for success in science. We now want to tie steps of critical thinking together with the scientific method so that you have a checklist for evaluating scientific experiments and papers.

In high school, you might have been asked to memorize the steps of the scientific method as if it were some magical list that would lead to experimental success. Although the scientific method (Box 3.3) is a key component of science, it is not a standard operating procedure for scientists. This is because it has to cope with many different situations—we can't run the "experiment" of creating the Universe again, for example.

BOX 3.3   Framework of the Scientific Method

1. Make observations.
2. Ask questions.
3. Generate a hypothesis and a null hypothesis.
4. Make predictions based on the hypothesis and null hypothesis.
5. Perform an experiment.
6. Analyze your results.
7. What new questions do you now have? Are more experiments necessary?

Sometimes the act of discovery alone, what you might think of as just the first step in the scientific method, is extremely important, and only later will scientists generate a hypothesis or test a theory in detail. But even in so-called discovery science, most of the time the scientists will have had to describe what they hope to see before being given access to experimental equipment to make observations or run experiments. You can think of that step as being similar to forming a question and creating a hypothesis, although it isn't usually phrased that way. Astronomy, in particular, is an example of a field where discovery science is very important.

Overall, it's best to think of the scientific method as a powerful tool that can be varied and applied in different ways by biologists, chemists, and physicists.

 Go to **science3.nelson.com** for examples of how the scientific method is applied in physics, chemistry, and biology.

The Scientific Method Consists of 10 Steps

I list the 10 steps of the scientific method in Box 3.4, and then I explain them in more detail below.

BOX 3.4   Critical Thinking and the Scientific Method Checklist

1. Become familiar with the question.
2. Assess the hypothesis.
3. Question how the facts were obtained.
4. Analyze the results.
5. Question the conclusion.
6. Ask if any assumptions were made.
7. Search for bias.
8. Evaluate the references and question the sources.
9. Question the relationships between the facts, the tiny details, and the big picture.
10. Leave emotion out of your analysis.

---

1. BECOME FAMILIAR WITH THE QUESTION. We talked earlier in this chapter about how scientists are curious about the world, mentioning that being curious involves asking questions. For any experiment, first become familiar with the specific question being asked. What observations led to the question? What background knowledge is the question based on? Make sure that you understand the question in detail.

2. ASSESS THE HYPOTHESIS. As you learned in Chapter 2, a **hypothesis** is a proposed explanation about the underlying cause for an observation. Some textbooks refer to a hypothesis as an "educated guess," but if you're guessing, you're probably not very clear in your hypothesis. When you undertake a scientific study, you might need to devise a hypothesis. (See Box 3.5 for a comparison of the terms *hypothesis* and *theory*.)

BOX 3.5   Hypothesis versus Theory

A *hypothesis* is a proposed explanation for a given observation. Note that a hypothesis can be general or specific.

A *theory* is an explanation for a large natural phenomenon and is very well supported by data. A theory is constructed over time by applying hypotheses systematically toward a natural phenomenon.

Recall from Chapter 2 that colloquial usage of the term *theory* is not scientifically correct. You have probably heard someone say, "Oh, that's just a theory." They're trying to tell you that there's a good chance what you're discussing is wrong. However, remember that a theory is a very well supported hypothesis (or collection of hypotheses) that is confirmed by a large amount of data.

Note that some textbook descriptions of the scientific method end with "formulating a theory." But this doesn't always happen. Depending on the type of science you are doing, it may be impossible to formulate a theory based on one experiment alone. You might need several to understand all the different factors that are present in the theory.

(We will discuss key theories of physics, chemistry, and biology in Chapter 6.)

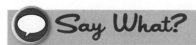

Say What?

HYPOTHESIS: The word *hypothesis* comes from the Greek term *hypothesis*, meaning "a supposition, something laid down as the foundation of an argument." Literally, *hypo* means "under" and *thesis* means "a placing."

NULL HYPOTHESIS: You will also see the term **null hypothesis**. The null hypothesis (also referred to as H_0) refers to a default position. For example, if the hypothesis is that a certain drug will provide a clinical benefit for a disease, then the null hypothesis is that the drug provides no benefit.

PREDICT: The word **predict** comes from the Latin verb *praedicere*, which means "to advise or foretell." If you break the word *praedicere* into *prae* (before) and *dicere* (to say), you see that its basic sense is "say before."

THEORY: The word **theory** comes from the Greek word *theoria*, meaning "contemplation," "speculation," or "theory."

In Box 3.6 I list the characteristics of a good hypothesis.

BOX 3.6   Characteristics of a Good Hypothesis (Hulley et al., 2001)

A good hypothesis

1. is simple,
2. is specific,
3. is plausible,
4. is stated in advance,
5. can generate predictions about new situations, and
6. is testable.

A good hypothesis is simple, specific, plausible, and stated in advance; can generate predictions about new situations; and is testable (Hulley et al., 2001). See Table 3.1 for examples of good and bad hypotheses.

Table 3.1 Good and Bad Hypotheses

Discipline	Good Hypothesis	Bad Hypothesis
Biology	Because this synthetic molecule is similar in shape to the natural hormone testosterone, we propose it can serve medically to compensate for patents with a naturally low level of testosterone.	Because this plant extract is naturally occurring, it won't have harmful side effects if you take it as a medicine.
Chemistry	Because molecule X has the same absorption pattern in the ultraviolet range as molecule Y, at least a portion of their structures must be identical.	Because molecule X has the same absorption pattern in the ultraviolet range as molecule Y, they must have identical structures.
Physics	The vector law of adding relative velocities applies only for speeds typically measured for projectiles in terrestrial laboratories.	Because the Sun glows in the sky, its primary power source is light.

Go to **science3.nelson.com** to access hypothesis-generating tutorials for physics, chemistry, and biology topics.

3. QUESTION HOW THE FACTS WERE OBTAINED. An experiment is performed to test the hypothesis, and you will want to pay special attention to the methods and **experimental design**. Box 3.7 outlines the characteristics of a well-designed experiment, and Box 3.8 provides a list of questions you should ask when critiquing an experiment.

BOX 3.7 Characteristics of Good Experimental Design

1. An appropriate control. If you're investigating a drug's efficacy, this might include the use of a placebo (which is a negative control).

2. Appropriate independent and dependent variables.

Continued

3. Adequate sample size. (This plays into the statistical significance of the work. More on that in Chapter 5.)

4. Minimization of confounding variables. Confounding variables are variables that the researcher didn't control or eliminate from the experimental design, and are different from the experimental variable.

5. Appropriate use of statistical analysis (again, more on that in Chapter 5).

6. Repeatability. The experiment was repeated. (Also called repetition.)

7. Randomization. Assignment to the control or experimental group was randomized.

8. Blindness. The experiment was blinded. This means that the experimenters were careful to eliminate their own biases when performing the experiment or interpreting the results.

 Go to **science3.nelson.com** for specific examples of good experimental design in physics, chemistry, and biology experiments.

BOX 3.8   Questions to Ask When Critiquing an Experiment

1. What was the control? Was it appropriate?

2. What were the independent variables? What were the dependent variables? Were they appropriate?

3. Are there replicates within the experiment? What is the sample size? Is the sample size appropriate?

4. Are there any confounding effects? How have confounding effects been minimized?

5. Were the correct statistical tests used?

6. Was the experiment repeated? How many times was it repeated? Was the experiment repeated by the same researchers or by other researchers who are independent of the first group (so if they find different results, they won't feel obliged to cover them up)?

7. Was there random assignment into the control and experimental groups? What was the method of randomization? Was the method adequate?

8. Is there any chance the experimenters were fooled by their results? For example, did they go out of their way to try to prove something they already believed in rather than do a "fair test" about whether they were correct in their reasoning?

Go to **science3.nelson.com** for specific examples of how to critique physics, chemistry, and biology experiments.

How to Critique Your Experiment

CONTROL. Experiments will usually have a control group and an experimental group. Both groups should be treated identically, except the experimental group is exposed to a given treatment, whereas the control group is not. Recall from Chapter 2 that the presence of a control makes the experiment more valuable and informative. Note that there are occasions where the inclusion of a control group is not possible.

VARIABLES. A variable is any part of the experiment that is subject to change. Recall that a good experiment minimizes any differences between the control group and the experimental group. When you design an experiment, make sure you consider all the dependent and independent variables. The *dependent variable* is what you're measuring. The *independent variable* is what you're manipulating. You might ask yourself what exactly the dependent variable "depends" on. The answer is the independent variable! (Or, at least, that is what you may be hypothesizing.) You'll see how to communicate these graphically in Chapter 8. The experiment is to see if there's a relationship between these, and it's this relationship that your hypothesis should be describing.

In Box 3.9, I describe examples of dependent and independent variables in physics, chemistry, and biology.

BOX 3.9   Examples of Dependent and Independent Variables

Biology example: When analyzing the effectiveness of a specific antibiotic on a bacterial infection, the independent variable is the presence of the antibiotic, and dependent variables could include the bacterial cell count in a test tube or detection of the bacteria in a patient sample.

Continued

Chemistry example: The concentration of a solution is the independent variable, if it is manipulated by you. The rate of a subsequent reaction is the dependent variable, because it depends on the concentration of the solution.

Physics example: If you are investigating Ohm's law for a resistor in an electrical circuit, the voltage applied is the independent variable, while the dependent variable is the current measured.

SAMPLE SIZE. As you will learn in Chapter 5, the size of the sample (or the *n* **number**) is linked to the statistical significance of the work. It is crucial that the experiment have an adequate **sample size** in order to minimize the effect that individual variability may have on your results.

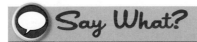

N NUMBER: The *n* refers to how many samples there were in an experiment. *n* is also called the sample size.

CONFOUNDING EFFECTS. It is important to attempt to eliminate **confounding effects** in experimental design. A confounding effect can occur when a factor that you are interested in studying is correlated with a different factor that you are not interested in or that is not being analyzed. The presence of confounding effects will disrupt and perhaps discredit any conclusions that could come from the data. You also need to be aware of **causation** versus **correlation** in analyzing your results. Take the graph in Figure 3.1, for example.

The graph in Figure 3.1 shows that Amazon.com sales have increased over a span of 10 years. The graph also shows that the number of vCJD cases has increased over a span of 10 years. (vCJD refers to variant Creutzfeldt-Jakob disease, a brain-wasting disease thought to be caused by the prion that causes mad cow disease.) Can you say from the graph that Amazon sales cause vCJD? The answer is of course "no." These two data sets are correlated

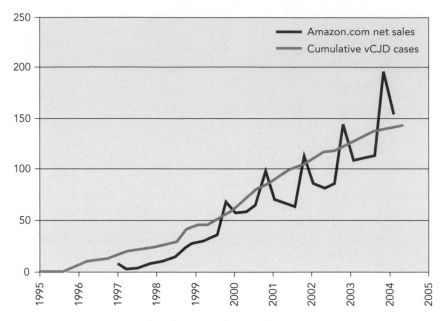

Figure 3.1 Do increased sales on the website Amazon.com cause increased cases of variant Creutzfeldt-Jakob disease (vCJD)?

Source: Bains, W. (2005). How to write up a hypothesis: The good, the bad, and the ugly. *Medical Hypotheses, 64*, 665–668.

(meaning they show the same trend in the same time period), but of course there is no causal link (one does not cause the other).

Another example that illustrates the difference between correlation and causation comes courtesy of blogger Bobby Henderson. Henderson satirically claims that dressing up as a pirate (such as during Halloween) causes the atmosphere to cool. Therefore, a good way to combat global warming is for more people to dress up like pirates. The evidence is stronger when you look back in time to when there were more pirates on Earth and extrapolate global temperatures to that time. This example clearly illustrates that correlation does not mean causation.

REPEATABILITY. Experiments are especially useful if it is possible to repeat them. This issue is made all the more important by the fact that one of the fundamental assumptions in scientific laws and theories is that they do not change with time. Some textbooks also speak of "generality" and the ability to extrapolate from one experiment to a more general condition. Upon critiquing an experiment, ask how many times the experiment was repeated and ask what larger concepts the results could be applied to.

RANDOMIZATION. Was **randomization** included in the experimental design? If so, how was it executed? It's worth emphasizing that randomization is more necessary in certain types of experiments than in others. In clinical trials, for example, randomization is a crucial component of experimental design. It is not enough to check if randomization was present in the experimental design. You must look deeper into the methods to see if the way randomization was implemented was adequate. Warning bells should go off in your head if a paper says that randomization occurred but neglects to indicate the methods.

BLIND EXPERIMENTAL DESIGN. This is typically most important when researchers are investigating a medical treatment, because they have the potential to be biased in their expectations, or are trying to measure something that's prone to being subjective. If an experiment is blind, then the subjects within the experiment are unaware of any treatment they may or may not be subjected to. If an experiment is double-blind, then both the subjects and the person running the experiment are unaware of which group (control or experimental) each participant is assigned to. Why is blinding important? Blinding can minimize the impact of the **placebo** effect (see Box 3.10) and reduce researcher bias in the results. The participants will likely exhibit differently if they know that they didn't get actual medication. Also, although the experimenters are genuinely interested in determining the results of the experiment, they are often not emotionless participants. They have chosen to study something because they have a connection to the outcome.

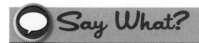

Say What?

PLACEBO: Since the late 1700s, the term *placebo* has been used to describe "a medicine given more to please than to benefit the patient," meaning that it will improve the patient's attitude about a given disease, rather than treat the disease itself. The word *placebo* comes from the Latin *placebo*, meaning "I shall please."

BOX 3.10   The Placebo Effect

The placebo effect is a documented and real phenomenon. It involves the improvement in a patient's condition when the patient is given a medicine they think to be real and beneficial, when in fact the medicine contains nothing that has been scientifically shown to benefit the patient in a pharmacologic manner. A common placebo is the sugar pill. There are other examples of the placebo effect outside of biology. For example, one study (Plassmann et al., 2008) used functional magnetic resonance imaging to scan people's brains when tasting wine. Before tasting each wine, the subject was told its price. The area of the brain associated with pleasure would light up more as the price of the wine increased. Even if the researchers gave the wine taster a bottle of relatively cheap wine, the brain scan would show more illumination of the "pleasure region" if tasters were told it was a more expensive wine.

Whenever you design or critique an experiment, the most important overall question to ask is, "Is this experiment a fair test?" In other words, does the experiment test what it says it tests, and does it test it well? Were any biases possible on the part of the experimenter or, if you're examining humans, the participants? After looking carefully at the details, step back and look at the experiment as a whole.

 Go to **science3.nelson.com** for examples of experimental design in physics, chemistry, and biology.

Sometimes it's impossible to achieve the "perfect" experimental design due to the various constraints of a given experiment or system. It is important that you familiarize yourself with the most common constraints (see Box 3.11), and with ways to alleviate or minimize their impact.

BOX 3.11   Potential Constraints on Good Experimental Design

1. Logistics
2. Cost
3. Available subjects
4. Time

Note that experiments are not the only means of testing a hypothesis, and all hypotheses are not tested in the same way. There are in fact situations where an experiment may be impossible. For example, in observational science it may not be possible to conduct a controlled experiment. Fields where observational science is common include paleontology, astronomy, and geology.

The Rest of the 10 Steps

4. ANALYZE THE RESULTS.   Carefully look at the results section and think about the analysis of the facts. Were statistics used? How were the experimental errors quantified? Were any fits to results made under reasonable assumptions such as the power law or exponential behaviour? Was the correct statistical test employed? Are the results significant? Students often get caught up in the equations and lose sight of what the numbers mean. This is what you have to focus on—the *meaning* of the numbers that are produced at the end of the analysis. As you will learn in Chapter 5, the study of fitting **data** and statistical analysis isn't scary. Statistics is, in fact, a wonderful tool and a crucial component of thinking critically about science. See Chapter 5 for a description of which statistical tests to use and when to use them.

When looking at the interpretation of results, look closely at the data and determine whether or not any data were excluded. If so, then ask why they were excluded. You should also take a good, hard look at the data themselves. Are any of the "data" anecdotal? (This will occur more frequently in magazine articles than in scientific papers.) Recall from Chapter 2 that **anecdotes** are stories, and remember that the "plural of anecdote is not data" (Ben Goldacre, 2011).

🗨 Say What?

DATA: *Data* is a Latin word that has been borrowed by English and means "things given": it comes ultimately from the verb *dare*, meaning "to give." It was used to describe storing computer information in the 1940s. Note that in English and Latin *data* is plural and *datum* is its singular.

ANECDOTE: Anecdote comes from the Greek word *anekdota*, which means "not" (*an*) "published" (*ekdota*).

5. QUESTION THE CONCLUSION. Look at the conclusion and question it. Is it supported by the results and statistical analysis? Could there be any alternative explanations? Note that it is possible that more than one hypothesis could be correct, especially if the experimental data are limited by the accuracy of your equipment. In fact, each hypothesis may explain some part of a larger observation or phenomenon.

 Go to **science3.nelson.com** for physics, chemistry, and biology examples of experimental outcomes that have alternative explanations.

6. IDENTIFY ASSUMPTIONS. Throughout the experiment, were any assumptions made? For example, were calibration levels or equipment accuracy assumed or checked carefully?

 Go to **science3.nelson.com** for physics, chemistry, and biology examples of common assumptions.

7. SEARCH FOR BIAS. Particularly in clinical trials it's important to ask who funded the study. What about the bias of the researchers? Are they trying to prove a theory that they came up with that they might want to be true? Everyone brings their own values to the table, so you need to make sure the researchers' biases aren't affecting their scientific reasoning about the problem at hand.

8. EVALUATE THE REFERENCES AND QUESTION SOURCES. Are the references peer-reviewed? (See Chapter 2 for the definition of *peer review*.) Are any key references omitted? You always need to question where the facts come from. This applies to your everyday (outside the science classroom) life as well. For example, salespeople at a store will probably use a narrower scope of facts and bring their own bias into the discussion.

9. QUESTION THE RELATIONSHIPS BETWEEN THE FACTS, THE TINY DETAILS, AND THE BIG PICTURE. Now it's time to put the topic into context. Can you extrapolate from the results? Think of both the small details and the larger picture when you do this. It's worth remembering that the smallest incorrect assumption about the behaviour of a piece of equipment can lead to an entire experiment producing corrupt data. At the same time, over-extrapolation is equally damaging. This is a major error of newspaper articles that claim a given compound will cure a particular disease. The journalist might jump from a drug showing a minimal change in an animal model to the drug having the capability to cure the disease in humans despite the complete lack of human clinical trial data.

10. LEAVE EMOTION OUT OF YOUR ANALYSIS. Is emotion involved? Does the person you are having a conversation with, or the advertisement you are looking at, or the scientific paper you are reading, try to appeal to your emotions? Note that you could naturally feel emotional towards an issue—but the question is whether the reporter is trying to make use of those emotions to influence your logic and reasoning. Make sure you leave emotion at the door. Emotion should not be a part of scientific reasoning. Accept what the data tell you even if it makes you uncomfortable. Sometimes that can be difficult, and it's OK to feel that, but you still need to leave emotion out of the equation.

 Go to **science3.nelson.com** for a profile of the autism and MMR scare and a breakdown of critical thinking.

3.5 CONCLUSION

Science is a process, not a destination. It's not a list of "facts" to obtain, but rather it's a systematic and creative way to investigate how aspects of the natural world interconnect. If the results of your experiment are not what you expect, that's OK! You can report data that do not fit a theory, provided you've done everything you can to figure out why, and have documented it. Each well-thought-out experiment leads to more study and better-refined hypotheses. If different tests of your hypothesis keep showing no connection between your dependent and independent variables, it's time to change how you think they might be interrelated—if indeed they are at all. Science is an ongoing process with no finish line.

> *"No single step in the pursuit of enlightenment should ever be considered sacred; only the search. ... "*

> ANN DRUYAN, *The Varieties of Scientific Experience*

FURTHER READING

Bains, W. (2005). How to write up a hypothesis: The good, the bad, and the ugly. *Medical Hypotheses, 64*, 665–668.

Chalmers, A. F. (1999). *What Is This Thing Called Science?* (3rd ed.). Buckingham, UK: Open University Press.

Goldacre, B. (2011). *Bad Science.* Toronto, ON: McClelland & Stewart.

Hollingworth, R. W., & McLoughlin, C. (2001). Developing science students' metacognitive problem solving skills online. *Australian Journal of Educational Technology, 17*, 50–63.

Hulley, S. B, Cummings, S. R., Browner, W. S., Grady, D., Hearst, N., & Newman, T. B. (2001). Getting ready to estimate sample size: Hypothesis and underlying principles. In *Designing Clinical Research: An Epidemiologic Approach* (2nd ed.), pp. 51–64. Philadelphia, PA: Lippincott Williams and Wilkins.

Kuhn, D. (1997). Constraints or guideposts? Developmental psychology and science education. *Review of Educational Psychology, 67*, 141–150.

McComas, W. F. (1996). Ten myths of science: Reexamining what we think we know about the nature of science. *School Science and Mathematics, 96*, 10–16.

Mpemba, E. B., & Osborne, D. G. (1969). Cool? *Physics Education, 4*, 172–175.

Northedge, A. (2003). Enabling participation in academic discourse. *Teaching in Higher Education, 8*(1), 169–180.

Plassman, H., O'Doherty, J., Shiv, B., & Rangel, A. (2008). Marketing actions can modulate neural representations of experienced pleasantness. *Proceedings of the National Academy of Science, 105*, 1051–1054.

Quinn, G. P., & Keough, M. J. (2002). *Experimental Design and Data Analysis for Biologists.* Cambridge, UK: Cambridge University Press.

Weinburgh, M. (2003). A leg (or three) to stand on. *Science and Children, 40*(6), 28–30.

Williams, W. M., Papierno, P. B., Makel, M. C., & Ceci, S. J. (2004). Thinking like a scientist about real world problems: The Cornell Institute for Research on Children Science Education Program. *Applied Developmental Psychology, 25*, 107–126.

Zimmerman, C. (2000) The development of scientific reasoning skills. *Developmental Review, 20*, 99–149.

Understanding Science

"If you can't explain something simply, you don't understand it well. Most of the fundamental ideas of science are essentially simple, and may, as a rule, be expressed in a language comprehensible to everyone. Everything should be as simple as it can be, yet no simpler."

ALBERT EINSTEIN

4

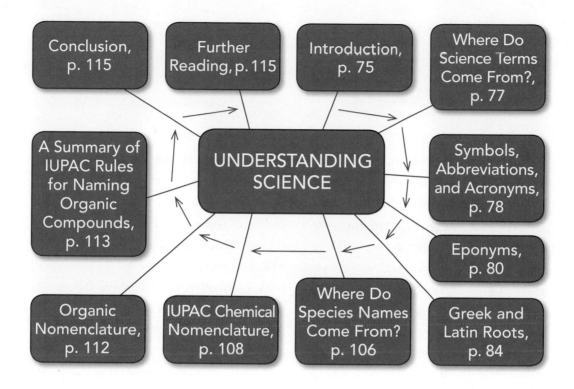

4.1 INTRODUCTION

If you feel that learning science is like learning a new language, you aren't alone. Many students have told me they have difficulty understanding sections of their textbook because of all the "science **jargon**." It's true that science, and each scientific discipline, has its own jargon. In order to understand science and succeed in your courses, you will have to learn this jargon. In order to learn a language, you need to learn the vocabulary. This can be a daunting task given that there are thousands of new scientific terms to learn.

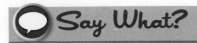
JARGON: The word derives from the Old French *gargun*, meaning the "cheeping of birds."

How will you go about studying these terms? Perhaps you should make lists of bolded words and their associated definitions, and then try to commit them to memory. Although this is a common study technique, it is not necessarily the most effective. Instead, it can be better to study the terms' **etymology**—where the words come from. If you study the source of a given word, and study the Greek and Latin roots that words are often based on (i.e., the **base word**), you will achieve a deeper level of understanding and be more able to recall them. An equally important benefit to this approach is that you will be able to dissect new words that you encounter in journal articles and figure out what they mean—even if it's the first time you've ever seen them!

Throughout the book so far I've discussed how science is about discovering things and making connections. If you want to make connections, it isn't enough just to know the definitions of words. You need to know where words come from and why things are given certain names. When or if you find yourself breaking new ground scientifically, you might have to communicate new observations or ideas in a way that makes the new information accessible to people who don't have your experience. Being able to create appropriate words to describe the events in your field is an important skill.

This is a key chapter of your book, and I expect that you'll access it frequently by referring to the tables of word roots and symbols as you come across new terms in your science classes. When you finish this chapter, you should understand the basis of scientific **terminology** and be familiar with keyword roots. Being conversant with proper scientific expression will make your job easier as you go through university. In fact, I view my classes as a conversation between me and my students. If you want to participate in this conversation, then you need to learn the language—the language of science.

DANGER ZONE

Many students spend their studying hours creating lists of bolded words and their associated definitions. Although this chapter contains a lot of lists, please remember that lists of words should not be where your studying ends. Knowing bolded words and definitions is not going to help you achieve success if you don't also know the context of scientific terms and understand their connections to scientific concepts. So as you read this chapter, try to focus on learning the root meaning (and where the word comes from) of the new words you see, rather than just remembering definitions. Also, note that participating in active conversations using terminology is a much better way to remember new words than just reading and reciting. (You'll learn more about active learning techniques in Chapter 7.)

4.2 WHERE DO SCIENCE TERMS COME FROM?
Eponyms and Descriptive Terms

There are two main ways that science concepts get names: (1) from an **eponym**, meaning it is named after something or someone (such as *Lou Gehrig's disease*), and (2) from a *descriptive term*, such as *leukocyte*, which comes from the Greek terms *leukos*, meaning "white," and *-cyte*, meaning "cell." (Note that *-cyte* ultimately comes from the Greek *kytos*, meaning "vessel.") This chapter has a section to discuss each of the above sources.

Note that throughout this chapter you'll see many new scientific terms. Please keep in mind that there are often two terms to describe the same thing: the common term and the scientific term. For example, Lou Gehrig's disease is the common name for *amyotrophic lateral sclerosis* (ALS), and white blood cell is the common name for *leukocyte*.

Before we discuss eponyms and descriptive terms, we'll take a look at symbols used in science. Before you can learn the words of a new language, you need to learn the alphabet, and symbols are like the alphabet of science.

4.3 SYMBOLS, ABBREVIATIONS, AND ACRONYMS

When learning a new language, you often start by studying some new vocabulary. However, before you can study this vocabulary, you need to learn the alphabet of the new language. Sometimes people compare learning the jargon of science to learning a new language, so we'll start by looking at symbols (the alphabet) that are used in science. Note that abbreviations and acronyms can also be considered symbols in science. Detailed lists of abbreviations used in physics, chemistry, and biology are located in Chapter 6 of this book.

The first type of symbols we'll look at are those that come from the Greek alphabet. Just as English letters have uppercase and lowercase forms, so do letters from the Greek alphabet (see Table 4.1). Letters from the Greek alphabet have been used consistently throughout all disciplines of science for particular concepts. For example, λ refers to the wavelength of a waveform in physics, the χ^2 test is a statistical test used in biology, a-helix and β-pleated sheet describe particular secondary structures found in proteins, and μ stands for the coefficient of friction in physics.

Table 4.1   The Greek Alphabet

Greek Alphabet Uppercase Letter	Greek Alphabet Lowercase Letter	English Spelling of Greek Letter
A	a	alpha
B	β	beta
Γ	γ	gamma
Δ	δ	delta
E	ε	epsilon
Z	ζ	zeta
H	η	eta
Θ	θ	theta

Continued

Greek Alphabet Uppercase Letter	Greek Alphabet Lowercase Letter	English Spelling of Greek Letter
I	ι	iota
K	κ	kappa
Λ	λ	lambda
M	μ	mu
N	ν	nu
Ξ	ξ	xi
O	o	omicron
Π	π	pi
P	ρ	rho
Σ	σ	sigma
T	τ	tau
Y	υ	upsilon
Φ	ψ	phi
X	χ	chi
Ψ	ψ	psi
Ω	ω	omega

Other symbols used in science, including many that are not derived from Greek, are shown in Table 4.2.

Go to **science3.nelson.com** for an expanded list of symbols used in science.

Table 4.2 Common Symbols Used in Science

Symbol	Meaning
Å	angstrom
f or v	frequency
∞	infinity
∅	empty set
℧	mho
∝	proportional to
☉	Sun
⊕	Earth
[ ]	concentration
!	factorial
Δ	change or heat
♂	male
♀	female
°	degree

4.4 EPONYMS

Many words, both common and scientific, are eponyms. An eponym is something (such as a person, place, or consumer item) that another thing is named after. For example, Parkinson's disease is named after James Parkinson, Mendelian inheritance is named after Gregor Mendel, Newtonian physics is named after Isaac Newton, and Ohm's law is named after Georg Ohm.

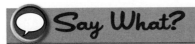
Say What?

EPONYM: The word *eponym* comes from the Greek word *eponymos*, which means "giving one's name to something."

Scientific terms that have names derived from eponyms are usually named for one of the following reasons: (1) for the person who discovered or invented it, (2) for the place it was discovered, (3) for the patient it was discovered in, (4) to honour a particular individual, or (5) because of a similarity in structure to another object. There are also many other reasons. Some examples of scientific phenomena or concepts that are derived from eponyms are illustrated below. The examples come from all three scientific disciplines: physics, chemistry, and biology.

Lou Gehrig's Disease

Lou Gehrig, an American baseball player, was diagnosed with amyotrophic lateral sclerosis (ALS) in 1939. ALS is a degenerative disorder characterized by progressive muscle atrophy and weakness, resulting in paralysis, the inability to eat, and the inability to breathe. Lou Gehrig's disease is the common name for this disorder, and amyotrophic lateral sclerosis is the scientific name. This is the disorder that afflicts Stephen Hawking.

Verneuil Process

Auguste Verneuil, a French chemist, announced in 1902 a "flame fusion" process that could be used to make corundum. Corundum is the mineral that rubies and sapphires are derived from, and the Verneuil process allowed rubies to be made relatively cheaply.

Becquerel

Antoine Henri Becquerel, a French scientist, along with Pierre and Marie Curie, discovered radioactivity and received a Nobel Prize in 1903. The becquerel (Bq) is an **SI (Système International)** unit of measurement for radioactivity.

Geiger Counter

In 1908, Hans Geiger, a German physicist, invented a machine that could detect radiation. The Geiger counter, although a different version, still bears his name today.

Golgi Apparatus

In 1897, Camillo Golgi, an Italian physician, first noted the existence of a particular membranous structure in eukaryotic cells. We now know that this membranous structure allows the cell to sort and modify proteins and lipids, kind of like an institution's "mail room."

Note that "Golgi apparatus" is one of the few organelle names that should be capitalized due to it referencing a proper name.

Hubble Space Telescope

Edwin Hubble, an American astronomer, discovered the existence of galaxies other than the Milky Way, gathered evidence in support of the theory that the universe is expanding, and established the basis for the Big Bang theory. The Hubble Space Telescope was named to honour him and was launched into Earth's orbit in 1990.

Heisenberg's Uncertainty Principle

Werner Heisenberg, a German physicist, was a pioneer in the development of quantum mechanics. As well as contributing important mathematical approaches, he was also able to show that you cannot simultaneously know both the position and the momentum of a quantum-mechanical system. The more you know about one of the two, the less you know about the other. This leads to strange phenomena, such as the Bose–Einstein condensate, in which super-cooled atoms (where the momentum is known to an extremely high degree of accuracy) tend to smear out to enormous sizes.

Schrödinger's Cat

This fictional character, often celebrated in literature and popular media, is part of a thought experiment proposed by Austrian physicist Erwin Schrödinger. The experiment is designed to show the incompatibility of an interpretation of quantum mechanics with the world in which we live. According to Schrödinger's interpretation of quantum mechanics, until a box containing a cat was opened, the cat would be a "living and dead cat (pardon the expression) mixed or smeared out in equal parts."

Richter Scale

In 1935, Charles Richter, an American physicist and seismologist, created a way to measure the size of earthquakes. The Richter magnitude scale is today known simply as the Richter scale.

Krebs Cycle

Hans Krebs, a British biochemist, identified the citric acid cycle, which is also known as the Krebs cycle, a discovery that earned him a Nobel Prize in Physiology or Medicine in 1953. Note that many students call the biochemical pathway "the Kreb's Cycle," but since his name is Krebs, the term is not a possessive.

Calvin Cycle

Melvin Calvin, an American chemist, characterized the Calvin cycle (C_3 cycle). This is the biosynthetic pathway that occurs in chloroplasts during photosynthesis to ultimately create complex organic molecules (such as glucose) from atmospheric carbon dioxide.

Down Syndrome

John Langdon Down, a British physician, wrote about Down syndrome in 1866. Note that Down syndrome is the common name. Trisomy 21 is the scientific name for this chromosomal disorder. Some people call this "Down's Syndrome," but that is not the correct term.

HeLa Cells

Henrietta Lacks was an American cancer patient from whom a sample of cancerous cervical cells was taken in 1951. HeLa cells (named for the first two letters in her first and last names) constitute an immortal cell line that is used in cancer research. It is currently the most commonly used cell line in scientific research and has been in continuous *in vitro* culture from the day it was first taken (Rahbari et al., 2009).

Alzheimer's Disease

In 1901, Alois Alzheimer, a German physician, published a description of "presenile dementia," a condition that is today known as Alzheimer's disease.

So how can you learn all the eponymous scientific terms (at least the ones you need to know for your courses)? There is no magic formula, and you will have to memorize them. However, memorization can become easier if you learn the history of how they were named. So, when you learn a new term in class, be sure to ask your instructor where that name comes from, or research it on your own or with friends outside of class. Talking about a new concept also helps you to appreciate its context and remember it.

 Go to **science3.nelson.com** for an expanded list of common eponymous scientific terms that you'll see in your first-year physics, chemistry, and biology classes.

4.5 GREEK AND LATIN ROOTS

The majority of scientific terms are descriptive, which means that they tell you something about the thing they are describing. Many of these descriptions come from Greek and Latin roots, and the words are made up of a **root** and/or **prefix** and/or **suffix**.

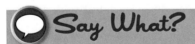

ROOT: A root (also called a **stem** or base) is the core part of a word that gives the word its central meaning, which can then be modified by the addition of a prefix and/or a suffix. For example, in the words *stability*, *stable*, and *destabilize*, the root is *stable*.

PREFIX: A prefix comes before the root of a word. For example, the word *replay* contains the prefix *re-* and the root *play*.

SUFFIX: A suffix (also called a **postfix**) comes after the root of a word. For example, the word *boyhood* contains the root *boy* and the suffix *-hood*.

Chances are good that you already know a large number of Greek and Latin roots/ prefixes/suffixes (in many cases without even realizing it!). What does *thermo-* (derived from Greek) mean? You probably recognize that term from the words *thermometer, thermal,* and *thermostat,* so you can probably deduce that *thermo-* means "heat." What does *hydro-* (derived from Greek) mean? You probably recognize that term from the words *fire hydrant, hydration,* and *hydraulic,* so you can probably figure out that *hydro-* means "water." Many of the Greek and Latin roots/prefixes/suffixes you know from everyday common language are also used in science jargon and terminology. For example, you'll recognize the Greek root *pod-* (the stem of the word for "foot," *pous*) from the words *tripod, podium,* and *octopus,* and you'll see it again in the terms *pseudopod* and *cephalopod* in your science courses. (A note about stems: I should tell you now that in Greek and Latin the endings of words usually change depending on their function in a sentence. The stem of a word is the basic form to which these changeable endings are added. Sometimes, a word will have more than one stem.)

Knowing some basic Greek and Latin terms will help you figure out the meaning of scientific words, even if you've never heard the words before. Let's take the example of a nudibranch. If your instructor says to you, "We'll discuss nudibranchs in class tomorrow," and you've never heard of them before, is there anything that you could learn about this animal from its name alone? Look closely at the word *nudibranch.* Do you recognize any words you've seen before? Perhaps you pull out the term *nudi.* Perhaps it reminds you of *nudist.* You're right—the *nudi* part of the word comes from Latin *nudus,* which means "naked." Next, you could pull out the term *branch.* What does that remind you of? Perhaps a tree branch, or perhaps the bronchioles of the lung. If you're thinking along the lines of "lung" then you're close. In this case, the word is derived from Greek *bragkhia,* which means "gills." So, the *nudibranch* animal's name means "naked gills"—and that's what it looks like—a sea slug with "naked" gills projecting from the outside of its body (see Figure 4.1).

 Go to **science3.nelson.com** for many more examples of interesting animal names, where Greek and Latin roots give you an idea about what the animal looks like.

Figure 4.1 The photo above is a nudibranch, a type of shell-less mollusc with external gills. Note that there are more than 3000 different species of nudibranchs.

Alexander Semenov/Science Photo Library

 STRATEGIES THAT WORK

Studying word roots, rather than just lists of definitions, will help you learn the words' true meaning and will also help you recall the terms long after you've taken a specific science course. Many instructors view learning etymology as a memory aid for remembering scientific terms, just as it is a tool for learning the English language (Hashemi & Aziznezhad, 2011).

Did you know that, in most first-year science courses, hundreds of words are listed in the glossary at the end of the textbook? Biology usually has the most terminology out of any science to learn, with the glossaries of first-year texts routinely containing more than a thousand terms. Many students come to me saying, "How am I supposed to memorize all this terminology?" My answer to them is to learn the key Latin and Greek roots. If you learn a handful of key roots, then you'll be able to decipher multitudes of scientific terms.

Also, learning the Greek and Latin roots will lead to greater conceptual understanding of scientific terms, as opposed to rote memorization and surface understanding. Once you learn a root, you will understand something about all words that use that root, and learning word roots is much easier than memorizing thousands of definitions. Don't get me wrong—you still need to study terminology. It's just that you should study smart (by learning word roots first) rather than just studying hard (writing down the definition of every single new word in your textbook, a list that would be immense). You'll see more tips about studying smart versus studying hard in Chapter 7.

Tables 4.3 and 4.4 outline the key Greek and Latin roots that you should know. Table 4.3 illustrates numerical roots, and Table 4.4 gives both informal and scientific examples of key Greek and Latin roots. As you go through the tables, try to think of other words you are familiar with that use the roots you see.

Table 4.3   Greek and Latin Numbers Used in Science

Number	Greek Prefix	Latin Prefix	Common Example
1	mono-	uni-	monogamy, uniform
2	di-/dy-	bi-/duo-	diurnal, biannual
3	tri-	tri-/terti-	triathlon, tertiary
4	tetra-	quadri-/quart-	tetrad, quadruple
5	penta-	quint-	pentagon, quintuple
6	hex-	sex(t)-, se-	hexagon, sextet
7	hepta-	sept-	heptanol, September
8	octa-	oct-	October
9	ennea-	non-/novem-	nonane, November
10	deca-	dec(a)-	December

Source: D.M. Ayers (1972). *Bioscientific terminology: Words from Latin and Greek stems.* Tucson, AZ: The University of Arizona Press; T.F. Hoad (Ed.). (1996). *Oxford concise dictionary of English etymology.* New York, NY: Oxford University Press.

Table 4.4   Common Greek and Latin Roots, Prefixes, and Suffixes in Science Terminology

Root, Prefix, or Suffix	Meaning(s)	Informal Example*	Science Example	Source†
a(n)-	without, not	atheist	anemic, asexual reproduction, anion	Greek a(n)-, meaning "not"
acro-	terminal, end, summit, tip	acronym	acroscopic, acrospore	Greek akros, meaning "terminal"
ad-	toward	address	adductor, adapt	Latin ad, meaning "toward"
aden(o)-	gland		adenoid	Greek aden, meaning "gland"
aero-	air, gas	aerodynamic	aerate, aerobic respiration	Greek aer, meaning "air"
agglutinat-	to glue together		agglutinate, agglutination	Latin agglutinare, meaning "to glue"
allo-	other, different	allophone	allogamy, alloplasty, allopatric speciation	Greek allos, meaning "other"
alveol-	hollow, cavity		alveolus	Latin alveus, meaning "concave vessel"
amphi-	around, both, two, dual	amphitheatre	amphipathic	Greek amphi, meaning "on both sides"
ana-	up, again, anew	analysis	anatomy, anabolic	Greek ana, meaning "up"
andr(o)-	male, man	philander	androgen, androgynous	Greek andro, meaning "man"
angi-	seed vessel		angiostomatous, gamentangium	Greek aggeion, meaning "vessel"
ann-	yearly, year	annual	perennial	Latin annus, meaning "year"

Continued

Root, Prefix, or Suffix	Meaning(s)	Informal Example[*]	Science Example	Source[†]
ante-	before	antecedent	antenatal	Latin *ante*, meaning "before"
anth(o)-	flower	anthology, chrysanthemum	anther, exanthema	Greek *anthos*, meaning "flower"
anthrop(o)-	man, human being	anthropology, philanthropy	anthropophilic	Greek *anthropos*, meaning "man"
anti-	against, opposite, instead of	antitrust	anticodon, antibody	Greek *anti*, meaning "against"
apic-	top, tip, summit, apex	apex	periapical, apical meristem	Latin *apic-*, the stem of *apex*, meaning "top"
apo-	away, off	apostrophe	apoptosis	Greek *apo*, meaning "away from, off"
aqu-	water	aqueduct	aquifer, aqueous humour	Latin *aqua*, meaning "water"
arachn-	spider	arachniphobia	Arachnida, arachnidium	Greek *arakhne*, meaning "spider"
arch(e)-	ancient, primitive, beginning	archaeology, archetype	adrenarche, archaea	Greek *arch-*, meaning "first, beginning"
arthr(o)-	jointed, articulated	arthritis	arthropod	Greek *arthron*, meaning "joint"
asc-	bag, sack		ascus, Ascomycetes	Greek *askos*, meaning "sack"
astr-, aster-	star	astrology, disaster	asteroid, Asteroidea	Greek *aster*, meaning "star"
atri-	room	atrium	atrium, sinoatrial	Latin *atrium*, meaning "court, hall, room"

Continued

Root, Prefix, or Suffix	Meaning(s)	Informal Example*	Science Example	Source†
auto-	same, self, spontaneous	autopsy, autograph, automobile	autism, autophagy	Greek *autos,* meaning "self"
aux(e)-	increase, grow, enlarge	auxiliary	auxin, auxochrome	Greek *auxein,* meaning "to increase"
bar-	weight, pressure	baritone	barometer	Greek *baros,* meaning "weight"
bene-	well	benefit	benign	Latin *bene,* meaning "well"
bi-	two	bicycle	binomial, bipolar	Latin *bi,* meaning "twice" or "two"
bio-	life, relating to living organisms	biology, biography	antibiotic, symbiosis	Greek *bios,* meaning "mode of life"
-blast	bud, sprout, germ, embryonic cell		astroblast, blastoderm	Greek *blastos,* meaning "bud, sprout"
brachi-	arm		brachiosaur	Greek *brakhion,* meaning "arm"
brachy-	short		brachysm	Greek *brakhys,* meaning "short"
bryo-	moss		bryophyte	Greek *bryon,* meaning "moss"
caps-	box	capsule	encapsulation	Latin *capsa,* meaning "box, chest"
carcino-	tumour	carcinogen	carcinoma	Greek *karkinos,* meaning "cancer"
cardio-	heart	cardiovascular, cardiac	cardioblast	Greek *kardia,* meaning "heart"

Continued

Root, Prefix, or Suffix	Meaning(s)	Informal Example*	Science Example	Source†
-carp, carp-	fruit		carpel, geocarpic	Greek *karpos*, meaning "fruit"
centr-	centre, middle	centrist	centripetal	Latin *centrum*, meaning "centre"
cephal-	head, brain		cephalopod, acrocephaly	Greek *kephale*, meaning "head"
chlor(o)-	green, chlorine	chlorine	chloroplast	Greek *khloros*, meaning "green/yellow colour"
chondr(o)-	cartilage, granule	hypochondriac	mitochondria, chondrocytes	Greek *khondros*, meaning "granule" or "cartilage"
chord-, -chord	cord, string	cord	chordate, notochord, urochord	Greek *khorde*, meaning "cord"
chromo-, chromat(o)-, -chrome	colour	monochrome	chromosome	Greek *khroma*, meaning "colour"
chym-, chyl-	juice		chyme, parenchyma	Greek *khymos* and *khylos*, both meaning "juice"
clad-	branch		clade, cladogram	Greek *klados*, meaning "branch"
clin-	to slope, to lean		syncline, thermocline	Greek *klinein*, meaning "to slope"
coll-	glue	protocol	collagen, colloid	Greek *kolla*, meaning "glue"
co(m)-, con-	together, with	cooperate	codominance, recombinant, condensation	Latin *co(m)-* and *con-*, meaning "together"

Continued

Root, Prefix, or Suffix	Meaning(s)	Informal Example*	Science Example	Source†
contra-	against, opposite	contradictory	contraception	Latin *contra*, meaning "against" or "opposite"
coron-	crown, top	crown	coronal	Latin *corona*, meaning "crown"
cosm(o)-	universe	cosmic	cosmological	Greek *kosmos*, meaning "universe"
crani-	cranium, the skull	cranium	amphicrania	Greek *kranion*, meaning "skull"
cri-, crin-, -crine	to separate, to secrete	critic, crisis	apocrine, endocrine	Greek *krinein*, meaning "separate"
crypt(o)-	hidden	cryptic, cryptography	cryptogenic, cryptozoic	Greek *kryptos*, meaning "hidden, secret"
cut-	skin	cuticle	intracutaneous, cutin	Latin *cutis*, meaning "skin"
cyst(o)-, cysti-, -cyst	bladder, cyst, sac	cystic fibrosis	acrocyst, hematocyst, oocyst	Greek *kystis*, meaning "bladder, pouch"
cyt-	cell		cytoplasm, syncytium	Greek *kytos*, meaning "vessel"
dem-	people, country	democracy, epidemic	pandemic	Greek *demos*, meaning "common people"
dendr-	tree	rhododendron	dendron, dendrite	Greek *dendron*, meaning "tree"
derm(o)-, dermat(o), -derm	skin	taxidermy	dermatophyte, mesoderm	Greek *derma*, meaning "skin"
deuter(o)-	second	Deuteronomy	deuterium	Greek *deuteros*, meaning "second"

Continued

Root, Prefix, or Suffix	Meaning(s)	Informal Example*	Science Example	Source†
di-	twice, double	diploma, dilemma	diploid, dioecious	Greek *di-*, meaning "two"
dino-	terrible	dinosaur	dinosaur	Greek *deinos*, meaning "terrible"
dis-	apart, away, separately	discard	dispersion	Latin *dis-*, meaning "apart"
dorm-	sleep	dormitory	dormant	Latin *dormire*, meaning "to sleep"
-duct	to lead, to carry	conduct, induce	abduction	Latin *ducere*, meaning "to lead, draw"
dynam(o)-	power	dynamite, dynamic	thermodynamics	Greek *dynamis*, meaning "force or power"
ecdys-	an escape, moulting		ecdysone	Greek *ekdyein*, meaning "to take off, remove"
eco-	house	economy	ecology, ectopic, ecosystem	Greek *oikos*, meaning "home, house"
ecto-	outside	ectopic pregnancy	ectotherm	Greek *ectos*, meaning "outside"
end(o)-	inside, within, inner	endogamy	endoplasmic reticulum	Greek *endon*, meaning "within"
enter-	intestine	gastroenterologist	parenteral	Greek *enteron*, meaning "intestine"
epi-	above, over	epilogue	epidermis	Greek *epi*, meaning "upon, at, before, near, on, over"
erg(o)-	work	energy, allergy	andrenergic	Greek *ergon*, meaning "work"

Continued

GREEK AND LATIN ROOTS

Root, Prefix, or Suffix	Meaning(s)	Informal Example*	Science Example	Source†
erythr(o)-	red		erythremia, erythrocyte	Greek *erythros*, meaning "red"
etho-	custom, habit		ethology	Greek *ethos*, meaning "custom, habit"
eu-	good, well, true		eukaryote	Greek *eu*, meaning "well"
ex(o)-	external, outside, out		exothermic, exergonic reaction	Greek *exo*, meaning "outside"
febr-	fever		febrile	Latin *febris*, meaning "fever"
ferment-	to decompose	ferment	fermentation	Latin *fermentare*, meaning "to ferment"
fibr-	fibre	fibre	fibrin, fibrinogen	Latin *fibra*, meaning "fibre"
fil-	thread	filagree	filament	Latin *filum*, meaning "thread"
fiss-	split		fission, fissure	Latin *fissus*, meaning "having been split"
flagell-	whip	flagellate	flagellum	Latin *flagellum*, meaning "whip"
flu-	flowing	flush	flux, fluid	Latin *fluere*, meaning "to flow"
gam-, -gamy, -gamous	marriage	monogamy, bigamy	gamete, syngamy	Greek *gamos*, meaning "marriage"
gastr(o)-	stomach, belly	gastronomic	gastropod	Greek *gaster*, meaning "stomach, belly"

Continued

Root, Prefix, or Suffix	Meaning(s)	Informal Example*	Science Example	Source†
gen(e)-	to produce	progenitor, degenerate	generation, genital	Latin *genus*, meaning "birth, origin"
geo-	Earth	geography	geodesy, geocentric	Greek *ge*, meaning "Earth"
glyc(o)-, gluc(o)-	sugar	glucose	glycogen, glycolysis	Greek *glykys*, meaning "sweet"
graph-, -graph, -graphy	to write	autograph, graphology	demography	Greek *graphein*, meaning "to write, scratch"
gymn(o)-	naked, uncovered	gymnasium	gymnosperm	Greek *gymnos*, meaning "naked"
hapl(o)-	single		haploid	Greek *haplous*, meaning "single"
helio-	Sun		heliocentric	Greek *helios*, meaning "Sun"
hemi-	half	hemisphere	hemibranch	Greek *hemi*, meaning "half"
hemo-	blood		hemophilia	Greek *haima*, meaning "blood"
hepat(o)-	liver	hepatitis	hepatocyte, heparin	Greek *hepar* (parallel stem *hepat-*), meaning "liver"
heter(o)-	different, other	heterogeneous	heterozygous	Greek *heteros*, meaning "the other of two, other, different"
hol(o)-	whole		hologram, holistic	Greek *holos*, meaning "whole"

Continued

Root, Prefix, or Suffix	Meaning(s)	Informal Example*	Science Example	Source†
homeo-, hom(o)-	likeness, similarity, same	homonym	homozygous	Greek *homoios,* and *homos,* meaning "same"
hydr(o)-	of water	hydrant, hydraulic	hydrophobic	Greek *hydro-,* from *hydor,* meaning "water"
hyp(o)-	below, under	hypotenuse	hypothermia	Greek *hypo,* meaning "below, under"
hyper-	above, over	hyperactive	hyperthermia	Greek *hyper,* meaning "above, over"
hyster-	later		hysteresis	Greek *husteros,* meaning "later"
infra-	below, under	infrastructure	infrared	Latin *infra,* meaning "below"
inter-	between	intersection	interspecies, interphase	Latin *inter,* meaning "between"
intra-, intro-	inside, within	introspective	intracellular	Latin *intra* and *intro,* both meaning "within"
is(o)-	equal, same	isosceles	isotonic, isotope	Greek *isos,* meaning "equal"
-it is	inflammation	arthritis	hepatitis	Greek *–itis,* meaning "pertaining to"
karyo-, caryo-	nucleus, nut, kernel		karyotype	Greek *karyon,* meaning "kernel, nut"
kine-, cine-, kinet-	to move	kinetic, cinema	kinesin, kinesiology	Greek *kinein,* meaning "to move"

Continued

Root, Prefix, or Suffix	Meaning(s)	Informal Example*	Science Example	Source†
lact(o)-	milk	lactose	lactation, lactase	Latin *lac*, meaning "milk"
lamin-	thin plate	laminate	lamella, nuclear lamina	Latin *lamina*, meaning "thin plate"
leuk(o)-, leuc(o)	white	leukemia	leukocyte	Greek *leukos*, meaning "white"
lip(o)-	fat, grease	liposuction	lipase	Greek *lipos*, meaning "fat"
lith(o)-, -lith	stone	lithography	lithosphere, megalith	Greek *lithos*, meaning "stone"
lum-	light	luminary	luminescence	Latin *lumen*, meaning "light"
-lyse, -lysis	break, loosen	analysis	lysis, hydrolysis	Greek *lysis*, meaning "dissolution"
macr(o)-	large	macroeconomics	macrogamy, macromolecules	Greek *macros*, meaning "long, large"
mal-	bad	malevolent	malignant, malpractice	Latin *malus*, meaning "bad"
mamm-	breast	mammogram	mammary gland	Latin *mamma*, meaning "breast"
meg(a)-	large	megaphone	megaspore, megalopic	Greek *megas*, meaning "great, large"
melan-	black	melancholy	melanin	Greek *melas*, meaning "black"
micr(o)-	small	microphone	microscope	Greek *micros*, meaning "small"

Continued

Root, Prefix, or Suffix	Meaning(s)	Informal Example*	Science Example	Source†
mito-	thread		mitochondria	Greek *mitos*, meaning "thread"
mon(o)-	one, single	monocle, monarch	monogeny	Greek *monos*, meaning "alone"
morph-, -morph, -morphous	form	morphine, amorphous	mesomorph	Greek *morphe*, meaning "form, shape"
mov-, mot-, mom-	move, motion	motor	momentum	Latin *movere*, meaning "to move"
mult(i)-	many	multitude	multiplex	Latin *multus*, meaning "many, much"
nephr(o)-	kidney		nephron	Greek *nephros*, meaning "kidney"
neur(o)-	nerve	neurotic	neuron	Greek *neuron*, meaning "nerve"
neutr(o)-	having no charge, neutral	neutral	neutron	Latin *neuter*, meaning "neither"
nihil-	nothing	nihilist	annihilation	Latin *nihil*, meaning "nothing"
nom-	name	nominate	binomial	Latin *nomen*, meaning "name"
noto-	back		notochord	Greek *noton*, meaning "back"
nucle(o)-	nucleus, nut	nuclear	nucleus	Latin *nucleus*, meaning "kernel, nut"
o(o)-	egg		oocyte, oogonium, oogenesis	Greek *oon*, meaning "egg"
od-	path, way		anode, diode	Greek *hodos*, meaning "a way"

Continued

Root, Prefix, or Suffix	Meaning(s)	Informal Example*	Science Example	Source†
olig(o)-	few, scant	oligopoly	oligandrous	Greek *oligos*, meaning "small, little, few"
onco-	tumour		oncology	Greek *ogkos*, meaning "bulk"
ortho-	straight, correct	orthotics	orthopedic	Greek *orthos*, meaning "straight, correct"
oste(o)-	bone	osteopath	osteoblast	Greek *osteon*, meaning "bone"
paed(o)-, ped(o)-	child, instruction	encyclopedia, pedagogical	pediatric	Greek *pais* (stem = *paid*-), meaning "child"
pale(o)-	ancient	paleography	paleobiology	Greek *palai*, meaning "long ago"
pan-	all	pantheism	pandemic	Greek *pant*, meaning "all"
par(a)-	beside, next to, near	parallel	parallax	Greek *para*, meaning "beside"
path(o)-, -path, -pathic, -pathy	disease, suffering	sympathy, pathetic	idiopathic	Greek *pathos*, meaning "feeling, suffering"
peri-	around, near	periphery	pericardium, perigee	Greek *peri*, meaning "around"
phag(o)-, -phage, -phagous, -phagy	to eat	sarcophagus	autophagy, phagocytosis	Greek *phagein*, meaning "to eat"
phen(o)-	appear		phenotype	Greek *phainein*, meaning "to show"

Continued

Root, Prefix, or Suffix	Meaning(s)	Informal Example*	Science Example	Source†
phil-, -phil, -phile, -philia, -philous	loving	philanthropy, philosophy	neutrophil, halophile	Greek *philos*, meaning "dear, beloved"
-phobe, -phobic	aversion to, fear	phobia	hydrophobic	Greek *phobos*, meaning "fear, panic"
phot(o)-	light	photograph	photosynthesis	Greek *phos* (parallel stem = *phot-*), meaning "light"
-phyll	leaf		chlorophyll	Greek *phyllon*, meaning "leaf"
phys-	nature	physical	physiology	Greek *phusis*, meaning "nature"
phyt(o)-	plant		phytoplankton	Greek *phuton*, meaning "plant"
pino-	drink		pinocytosis	Greek *pinein*, meaning "to drink"
-plast	something moulded, formed	plastic, plaster	chloroplast, plastid	Greek *plastos*, meaning "formed"
plec-	interwoven		symplectic	Greek *pleckein*, meaning "to twine, plait, weave"
pluri-	more, many	plural	pluriparity	Latin *pluri-*, from *plus*, meaning "more"
-pod, -podium, podo-	foot	tripod, podium, octopus	podiatrist, pseudopodia	Greek *pous* (stem = *pod-*), meaning "foot"

Continued

Root, Prefix, or Suffix	Meaning(s)	Informal Example*	Science Example	Source†
poly-	many	polygon	polymer	Greek *polys*, meaning "many"
-port	to carry	transport	transporter	Latin *portare*, meaning "to carry"
post-	after, behind	postscript	postsynaptic cell	Latin *post*, meaning "behind, after"
pre-, pro-	before	pretest	precambium, prokaryote, promoter	Latin *prae* and *pro*, meaning "in front, before"
prot(o)-	first, original	prototype, protagonist	protozoan	Greek *protos*, meaning "first"
pseud(o)-	false	pseudonym	pseudopod, pseudogene	Greek *pseudes*, meaning "false"
psych(o)-	spirit, soul	psyche	psychology	Greek *psykhe*, meaning "soul"
quadr-, quadru-	four	quadruple	quadruped	Latin *quadri-* and *quadru-*, meaning "four"
quant-	how much	quantity	quantifiable	Latin *quantus*, meaning "how much"
quin-	five, fifth	quintuple	quintile	Latin *quintus*, meaning "fifth"
radi-	spoke of a wheel, ray	radiator, radio	radiole, radiate	Latin *radius*, meaning "ray"
recti-	straight	rectify	rectilinear	Latin *rectus*, meaning "straight"
retro-	backward, behind	retroactive	retrograde	Latin *retro*, meaning "backward"

Continued

Root, Prefix, or Suffix	Meaning(s)	Informal Example*	Science Example	Source†
rhiz(o)-, -rhiza	root	rhizome	mycorrhiza	Greek *rhiza*, meaning "root"
sarco-	flesh	sarcophagus	osteosarcoma	Greek *sarx*, meaning "flesh"
saur(o)-, -saur	lizard		dinosaur, branchiosaur	Greek *sauros*, meaning "lizard"
-scope	observe, view	scope	telescope, microscope	Greek *skopein*, meaning "to observe"
sec-, seg-	cut	segment	secant	Latin *secare*, meaning "to cut"
semi-	half, partial	semicircle	semicircular canals	Latin *semi-*, meaning "half"
sider-	star		sidereal	Latin *sidus* (stem = *sider-*), meaning "star"
solv-	to loosen, dissolve	insolvent	solute, solvent	Latin *solvere*, meaning "to loosen"
som-, -some	body		dermatosome, somite, acrosome	Greek *soma*, meaning "body"
sperm-, -sperm, spermat(o)-	semen, seed	sperm	angiosperm, spermatheca	Greek *sperma*, meaning "seed"
spor(o)-	seed	sporadic, diaspora	sporont, acrospore	Greek *spora*, meaning "seed, a sowing"
sta-	to stand, to stop, to fix	static, thermostat	hemostasia, epistasis	Greek *histemi* (stem = *sta-*), meaning "to make stand"

Continued

Root, Prefix, or Suffix	Meaning(s)	Informal Example*	Science Example	Source†
stoma-	mouth, opening	stomach	deuterostome	Greek *stoma*, meaning "mouth"
sym-, syn-	together, with	symphony	symbiosis, synapse, sympatry	Greek *syn*, meaning "together, with"
tach-	swift	tachometer	tachycardia	Greek *tachus*, meaning "swift," and *tachos*, meaning "speed"
taxo-	to arrange, to put in order	tactics, syntax	taxonomy, taxon, phototaxis	Greek *taxis*, meaning "arrangement"
tel(e)-	afar	television, telephone	telescope	Greek *tele*, meaning "afar"
telo-	end		telomerase, telomere	Greek *telos*, meaning "end"
tetra-	four	tetrahedron	tetrapod	Greek *tettares*, meaning "four"
therm(o)-	heat	thermometer, thermostat	hypothermia	Greek *therme*, meaning "heat"
thyro-, thyreo-	shield		thyroid	Greek *thyreos*, meaning "shield"
tom-, -tomy	to cut, section	anatomy, atom	appendectomy	Greek *tome*, meaning "a cutting"
top(o)-, -tope	place	topical	isotope, epitope	Greek *topos*, meaning "place"
trans-	across, through	transport	transporter, transcription, translation	Latin *trans*, meaning "across, through"
tri-	three	triple	trimester	Greek *tri-*, meaning "three"

Continued

GREEK AND LATIN ROOTS **103**

Root, Prefix, or Suffix	Meaning(s)	Informal Example*	Science Example	Source†
trop-	a turn	tropic	phototropic	Greek *tropos*, meaning "a turn, direction, way"
-troph	nourishment, development		heterotroph, autotroph	Greek *trophe*, meaning "food, nourishment"
ultra-	beyond	ultralight	ultrasonic	Latin *ultra*, meaning "beyond"
umbr-	shade, shadow	umbrella	penumbra	Latin *umbra*, meaning "shade"
uni-	one	unique, uniform	uniporter, uniparous	Latin *uni-*, meaning "one"
vacu-	empty	vacuum	vacuole	Latin *vacuus*, meaning "empty"
val-	to be strong, to be well	valid, valedictorian, evaluate	valence, covalent bond, bivalent	Latin *valere*, meaning "to be strong"
vas-	vessel	vase	vascular, cardiovascular	Latin *vas*, meaning "vessel"
verber-	strike, beat		reverberation	Latin *verberare*, meaning "to strike"
visc-	thick		viscosity	Latin *viscum*, meaning "a sticky substance"
viv-	living	survive, vivid, vivacious	viviparous	Latin *vivus*, meaning "living"

Continued

Root, Prefix, or Suffix	Meaning(s)	Informal Example[*]	Science Example	Source[†]
volv-, volut-	roll	revolve	convolution	Latin *volvere*, meaning "to turn"
vom-	discharge	vomit	vomit	Latin *vomere*, meaning "to discharge, vomit"
xen(o)-	foreign	xenophobia	xenobiotic	Greek *xenos*, meaning "foreign"
xyl(o)-	wood	xylophone	xylem	Greek *xylon*, meaning "wood"
zo-, -zoa(n), -zoic, -zoon	animal, living being	zodiac	metazoan	Greek *zoe*, meaning "life," and *zoon*, meaning "living being"
zyg(o)-	yoke		zygote, heterozygous, monozygotic	Greek *zygon*, meaning "yoke"

[*]Not all roots, prefixes, and suffixes have an informal example.
[†]Information for these terms came from these sources:

D.M. Ayers (1972). *Bioscientific Terminology: Words from Latin and Greek Stems.* Tucson, AZ: The University of Arizona Press.

For a searchable etymological dictionary of science terms, go to **science3.nelson.com**.

4.6 WHERE DO SPECIES NAMES COME FROM?

"Taxonomy is described sometimes as a science and sometimes as an art, but really it's a battleground."

BILL BRYSON, *A Short History of Nearly Everything*

Taxonomy is how we define and group species based on similarities and shared common ancestry. The groups are arranged in a hierarchical order, or "taxonomic ranks," moving from most inclusive to least inclusive. The names of the taxonomic ranks are domain, kingdom, phylum, class, order, family, genus, and species. An easy way to remember this is to use the mnemonic **d**o **k**eep **p**ots **c**lean **o**r **f**amily **g**ets **s**ick, or **d**o **k**ids **p**refer **c**heese **o**ver **f**ried **g**reen **s**pinach.

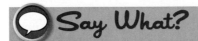

Say What?

TAXONOMY: This word comes from the Greek words *taxis*, meaning "arrangement," and -*nomia*, meaning "management."

In this section, we'll discuss where species get their names. When discussing a species, we could use the common name (such as killer whale) or the scientific name (such as *Orcinus orca*). The scientific name consists of two words: The first word is capitalized and refers to the genus; the second word refers to the species. Both words are either *italicized* or <u>underlined</u> (underline if you're writing them down during a test!). When a new species is discovered, the scientist (or scientists) who discovers it gets to name it. A specific name may be given to a species for widely varying reasons, some of which are (1) a distinguishing structure or feature of the species, (2) where it was found, (3) mythology, (4) the scientist(s) who discovered it, (5) the name of a person whom the scientist(s) wanted to honour, or (6) an acronym.

Read on for a few examples that depict the diverse nature (usually serious, but sometimes comical) of species naming.

Species Named After Commercial Products

Orizabus botox This species of scarab beetle was given its name because it is very smooth (hence the term *botox* in the name) (Ratcliffe & Cave, 2006).

Agave tequilana This species of agave plant is used to make tequila (hence the word tequila in the name).

A Species Named After an Acronym

Melanotaenia angfa This species of rainbowfish has *angfa* as a species name because it was named in honour of the Australia New Guinea Fishes Association.

Species Named Based on Their Location

Amblyoproctus boondocksius This species of scarab beetle is found "in the middle of nowhere" and thus has the term *boondocks* in its name (Ratcliffe, 1988).
Haitia This is a genus of plant found in Haiti.

Species Named After Celebrities

Agra schwarzeneggeri The males of this species of carbid beetle have a very developed middle femora (part of the arthropod leg), resulting in the appearance of biceplike bumps (hence the use of *schwarzenegger* in the species name) (Erwin, 2002).
Norasuphus monroeae This trilobite is shaped like an hourglass (hence the use of the name *monroeae,* after Marilyn Monroe) (Fortey & Shergold, 1984).
Aptostichus angelinajolieae This species of trapdoor spider is named after actress Angelina Jolie, though it isn't known exactly why the scientist chose this name (Bond & Stockman, 2008).

Species That Were Misnamed

These species still bear their original names, but the names were given to them for mistaken reasons.

Arctocephalos pusillus This species of seal is called *pusillus,* which means "very small." However, the adult seal is in fact quite large—weighing 1 tonne. It was given its name based on the analysis of a juvenile, rather than adult, organism.
Haemophilus influenzae This bacterium was incorrectly named *influenzae* because it was thought to be the cause of flu in the late 1800s and early 1900s.

4.7 IUPAC CHEMICAL NOMENCLATURE

The two basic areas in chemical nomenclature are inorganic nomenclature and organic nomenclature. You were probably introduced to the basics of both in high school. While most compounds have common names, it is best to learn the systematic names. There are several different naming systems, but I will be using the International Union of Pure and Applied Chemistry, or **IUPAC**, system. This will serve as a bit of a review and perhaps a preview of material you will learn (or have learned) in first- and second-year university chemistry classes.

Inorganic Nomenclature

It is important that every single inorganic compound have a name from which you can deduce its molecular formula. In order to achieve this, inorganic chemical compounds are named using the IUPAC systematic method of nomenclature. There are, however, some compounds for which the systematic name is never used except when someone is telling a really lame science joke. These are H_2O (water), H_2O_2 (hydrogen peroxide), and NH_3 (ammonia).

Naming Ionic Compounds

The rules in their most reduced form are given below:

1. The positive portion (metal cation, positive polyatomic ion, hydrogen ion, or less electronegative non-metal) is named and written first.

2. The negative portion (more electronegative non-metal or negative polyatomic ion) is named and written last.

3. For all binary compounds (compounds containing two different elements), the ending of the second element is -*ide*.

   a. When both elements are non-metals, the number of atoms of each element is indicated in the name with Greek prefixes. When no prefix appears for the first element, one atom is assumed. (See Box 4.1 for the prefixes.)

   Examples:

      S_2F_6: disulfur hexafluoride SiO_2: silicon dioxide

   b. When the positive part is a metal with only one oxidation state (i.e., group IA or IIA, or Al, Cd, Zn, or Ag), no Greek prefixes are used. The names are given using rules 1 and 2 above.

   Examples:

      NaCl: sodium chloride Al_2O_3: aluminum oxide

   c. When the positive part is a metal other than those named in 3b above, the charge is specified by a Roman numeral in parentheses, immediately following the name of the metal. The negative part is named as in rule 2 above.

   Examples:

      $FeCl_2$: iron (II) chloride Hg_2Br_2: mercury (I) bromide

4. For all ternary or higher compounds (compounds containing three or more different elements), the same procedure is followed as for binary compounds except that the name of the polyatomic ion is included. See Box 4.2 for the names of some common polyatomic ions.

   Examples:

      $CoPO_4$: cobalt (III) phosphate $Zn(OH)_2$: zinc hydroxide

5. Compounds with water molecules bound to the salt are referred to as hydrates. Such a compound is named as usual following the rules above, and then a Greek prefix (indicating the number of water molecules) followed by the root -*hydrate* is added to the name.

   Example:

      $Na_2CO_3 \cdot 10H_2O$: sodium carbonate decahydrate

1	mono-	6	hexa-
2	di-	7	hepta-
3	tri-	8	octa-
4	tetra-	9	nona-
5	penta-	10	deca-

BOX 4.2   Common Polyatomic Ions (excluding those given previously)

Name	Formula	Name	Formula	Name	Formula
acetate	$C_2H_3O_2^-$	dichromate	$Cr_2O_7^{2-}$	peroxide	O_2^{2-}
ammonium	NH_4^+	hydrogen carbonate	HCO_3^-	phosphate	PO_4^{3-}
bromate	BrO_3^-	hydroxide	OH^-	oxalate	$C_2O_4^{2-}$
carbonate	CO_3^{2-}	nitrate	NO_3^-	sulfate	SO_4^{2-}
chromate	CrO_4^{2-}	nitrite	NO_2^-	sulfite	SO_3^{2-}
cyanide	CN^-	permanganate	MnO_4^-	thiocyanate	SCN^-

Naming Hydrates

Hydrates are ionic compounds with absorbed water surrounding them. They are named as the ionic compound followed by a numerical prefix and the root -hydrate. The numerical prefixes used are listed in Box 4.1 above.

Naming Inorganic Acids

Compounds that, when dissolved in water, liberate hydrogen ions are often referred to as acids. A binary acid (a compound composed of hydrogen and a non-metallic element) in its pure state is named following the rules given above. However, once that acid is dissolved in water, the solution is given a different name. HCl is a simple example. In the gas phase, HCl is named hydrogen chloride, but when it is dissolved in water, it becomes hydrochloric acid.

The other common type of inorganic acid is the oxy acids. These compounds contain three elements—hydrogen; oxygen; and another element, such as chlorine or phosphorus. (See Box 4.3 for examples of some of the other elements that might combine with hydrogen and oxygen to form an oxy acid.)

The rules for naming acids, in their most reduced form, are given below:

1. Determine if the compound contains oxygen.

2. If the anion DOES NOT CONTAIN OXYGEN, then name the acid with the prefix *hydro-* and the suffix *-ic* attached to the root name of the non-hydrogen element. Follow this with the word *acid.*

3. If the anion DOES CONTAIN OXYGEN, then name the acid using the root name of the central element of the anion (the non-hydrogen, non-oxygen element) or use the name of the anion followed by the suffix *-ic* or *-ous.* The name is finished by adding the word *acid.*

   a. If the name of the anion ends with *-ate,* use the *-ic* suffix.

   b. If the name of the anion ends with *-ite,* use the *-ous* suffix.

   c. This can be little more complicated—please see Box 4.4 for greater detail.

BOX 4.3	Some Common Inorganic Acids (all names are followed by the word acid)				
Carbonic	H_2CO_3	Hydrochloric	HCl	Nitrous	HNO_2
Chloric	$HClO_3$	Hydrocyanic	HCN	Perchloric	$HClO_4$
Chlorous	$HClO_2$	Hydroiodic	HI	Phosphoric	H_3PO_4
Hydrobromic	HBr	Nitric	HNO_3	Sulfuric	H_2SO_4

When an element can combine with oxygen to form more than one oxyanion, use the following rules to name them. The one with the most oxygen atoms gets the -*ate* ending, and the other one, with fewer oxygen atoms, gets the -*ite* ending. Some of the halogens can combine with oxygen in four different ways. The anion with an extra oxygen is named *per____ate*. The anion with the fewest oxygens is named *hypo___ite*. For chlorine, the series is perchlorate (ClO_4^-), chlorate (ClO_3^-), chlorite (ClO_2^-), and hypochlorite (ClO^-).

4.8 ORGANIC NOMENCLATURE

When naming organic compounds according to the IUPAC system, it is important to keep in mind that every name consists of at least four parts: numbers, prefixes, roots, and suffixes. What these parts tell you are the locations and identities of the substituents and the parent compound to which the substituents are attached.

The easiest way to learn organic chemistry naming rules is to start with the easiest group of compounds, the alkanes. Alkanes are the simplest of a larger group of compounds known as hydrocarbons, which contain only hydrogen and carbon. Alkanes have only single bonds and are sometimes referred to as being saturated because no multiple bonds are present. Box 4.5 contains the names of the alkanes made from one, two, three, and four carbons. Also shown are the names of the substituents that can be derived from these alkanes. The words in italics indicate common names where they are different from the IUPAC name.

Box 4.6 on page 114 contains the names of the alkanes corresponding to C_1 to C_{10}. The names are for what are referred to as straight-chain or normal alkane; they are called this because they contain no branches or substituents.

BOX 4.5 Development of Ideas for Naming

		Alkane − H = alkyl group				
Molecular Formula	Alkane Structure	Alkane Name	− H	=	Alkyl Group Structure	Alkyl Group Name
CH_4	H H—C—H H	Methane	− H	=	CH_3—	Methyl
C_2H_6	CH_3CH_3	Ethane	− H	=	CH_3CH_2—	Ethyl
C_3H_8	⋀	Propane	− H	=	$CH_3CH_2CH_2$ — CH_3CHCH_3	Propyl *Isopropyl* (1-methylethyl)
C_4H_{10}	⋀⋁	Butane	− H	=	$CH_3CH_2CH_2CH_2$— $CH_3CH_2CHCH_3$	Butyl *Sec-butyl* (1-methylpropyl)
	⋏ (branched)	*Isobutane* 2-methyl-propane	− H	=	CH_3 CH_3CHCH_2— CH_3 CH_3CCH_3	*Isobutyl* (2-methylpropyl) *tert-butyl* (1,1-dimethylethyl)

4.9 A SUMMARY OF IUPAC RULES FOR NAMING ORGANIC COMPOUNDS

We will review these rules by working through an example:

$$CH_3 \qquad\qquad CH_2CH_3$$
$$CH_3CHCHCH_2CH_2CH_2CHCH_3$$
$$CH_3$$

BOX 4.6 Names of the First 10 Normal Alkanes

1 carbon—methane	6 carbons—hexane
2 carbons—ethane	7 carbons—heptane
3 carbons—propane	8 carbons—octane
4 carbons—butane	9 carbons—nonane
5 carbons—pentane	10 carbons—decane

1. Find the longest continuous chain. (In Example 1 it is a nine-carbon chain because you can go around bends and still be continuous.)

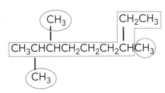

2. Name the parent compound. (Refer to Box 4.6 to find the name for a nine-carbon chain.)

3. Find and label the substituents. (If you put a line through the nine carbons you have chosen for the parent, whatever is left hanging out is a substituent. In this case you have three methyl groups hanging off the main chain.)

$$CH_3CHCHCH_2CH_2CH_2CHCH_3$$

4. Use a prefix to designate repetitions of a substituent, that is, 2 is di, 3 is tri, 4 is tetra, and so on. (In this case, there are three substituents, so you will call it trimethyl.)

5. Give each substituent the lowest number possible; that is, start at the end that gives the lower number at the first point of difference. (If you follow the chain from the left-hand side, the first methyl group is on carbon 2. If you follow the chain from the right-hand side, the first methyl group is on carbon 3. This means that numbering from the left is preferred in this molecule, so the methyl groups are at carbons 2, 3, and 7.)

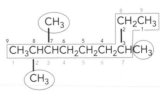

6. In the name there should be no gaps between the words (unless there is a number).

7. Numbers are separated by a comma.

8. Letters and numbers are separated by a hyphen.

Putting the name of the compound together we get 2,3,7-trimethylnonane.

 Go to **science3.nelson.com** for more information on organic nomenclature. For example, you'll see how to name cyclic alkanes, alcohols, and other functional groups.

4.10 CONCLUSION

"It is impossible to disassociate language from science. … To call forth a concept, a word is needed."

ANTOINE LAVOISIER (generally considered to be the father of modern chemistry)

Upon finishing this chapter, you should understand the basis of scientific terminology; be familiar with keyword roots; and have an awareness of where the language of physics, chemistry, and biology comes from. This will allow you to engage in conversations about scientific topics with your instructor and your peers. Whatever science you are studying, you must learn the sublanguage behind it; learning the sublanguage well will lead to a better and easier understanding of the science.

This chapter has introduced a lot of scientific words. Words are simply tools used to describe things and concepts. Numbers are also tools, and in the next chapter we'll explore the basic mathematical concepts that permeate science.

FURTHER READING

Ayers, D. M. (1972). *Bioscientific Terminology: Words from Latin and Greek Stems.* Tucson, AZ: The University of Arizona Press.

Bond, J. E., & Stockman, A. K. (2008). An integrative method for delimiting cohesion species: Finding the population–species interface in a group of Californian trapdoor spiders with extreme genetic divergence and geographic structuring. *Systematic Biology, 57*(4), 628–646.

Bryson, B. (2004). *A Short History of Nearly Everything.* New York, NY: Broadway Books.

Erwin, T. L. (2002). The beetle family Carabidae of Costa Rica: Twenty-nine new species of *Agra fabricius* 1801 (Coleoptera: Carabidae, Lebiini, Agrina). *Zootaxa, 119,* 1–68.

Fortey, R. A., & Shergold, J. H. (1984). Early Ordovician trilobites, Nora Formation, Central Australia. *Palaeontology, 27,* 315–366.

Glare, P. G. W. (Ed.). (1982). *Oxford Latin Dictionary.* New York, NY: Oxford University Press.

Hashemi, M., & Aziznezhad, M. (2011). Etymology: A word attack strategy for learning the English vocabulary. *Procedia—Social and Behavioral Sciences, 28,* 102–106.

Hoad, T. F. (Ed.). (1996). *Oxford Concise Dictionary of English Etymology.* New York, NY: Oxford University Press.

Liddell, H., Scott, R., & Jones, H. (Eds.). (1996). *A Greek–English Lexicon* (Rev. ed.). New York, NY: Oxford University Press.

Oxford English Dictionary. (2012). http://www.oed.com. Oxford University Press.

Rahbari, R., Sheahan, T., Modes, V., Collier, P., Macfarlane, C., & Badge, R. M. (2009). A novel L1 retrotransposon marker for HeLa cell line identification. *BioTechniques, 46*(4), 277–284.

Ratcliffe, B. C. (1988). New species and distributions of neotropical phileurin and a new phileurine from Burma (Coleoptera: Scarabaeidae: dynastinae). *Papers in Entomology,* Paper 73.

Ratcliffe, B. C., & Cave, R. D. (2006). The dynastine scarab beetles of Honduras, Nicaragua and El Salvador. *Bulletin of the University of Nebraska State Museum, 21,* 1–424.

Science and Math

"The essence of mathematics is not to make simple things complicated, but to make complicated things simple."

STAN GUDDER, Professor of Mathematics, University of Denver

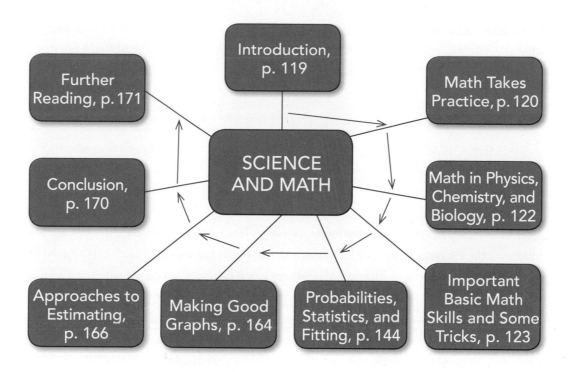

Introduction,
p. 119

Further
Reading, p. 171

Math Takes
Practice, p. 120

SCIENCE
AND MATH

Conclusion,
p. 170

Math in Physics,
Chemistry, and
Biology, p. 122

Approaches to
Estimating,
p. 166

Making Good
Graphs, p. 164

Probabilities,
Statistics, and
Fitting, p. 144

Important
Basic Math
Skills and Some
Tricks, p. 123

5.1 INTRODUCTION

Believe it or not, math is your friend. It's one of the few things invented by humanity that can make complicated things much easier. But it can also be intimidating for some—there is only one right solution or set of solutions for a given problem in **mathematics**. There are no opinions or points of view. So you need to know your stuff! But the good news is, armed with a few basic skills, most students can master the mathematics needed for a major in the sciences. And if you really enjoy math (that's great), you'll probably find it easier to learn!

It's hard to understate how incredibly important math is. Everybody knows that without an understanding of advanced math even the most simple of modern technologies wouldn't be possible. But perhaps even more amazing is that math seems to be incredibly good at describing the Universe. From the behaviour of the smallest subatomic particles we know, to the structure of the entire Universe, math can describe things that are pretty much impossible to do with words alone. It's difficult to imagine what science would be like without it.

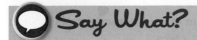

Say What?

MATHEMATICS: The word *mathematics* comes from the Greek word *mathema*, which means "what you know" or "what you learn."

Math also comes in a lot of different forms. While we learn about sets, arithmetic, and algebra in school, there's actually a whole lot more math out there. For example, math can be used to study properties of surfaces (the field of **topology**), or it can help you decide whether a logical question has an easy or complex answer (computational complexity theory). As math gets harder, it also tends to look less and less like basic arithmetic and starts to rely upon different kinds of logic. Unless you take a math degree you'll probably never study these particular fields, but it's good to know they are out there and when they might become useful.

The sections in this chapter review, and perhaps in some cases introduce, a number of important mathematical concepts. I want to emphasize that while this chapter is designed to highlight some key math skills, it can't possibly cover everything you will need in a science degree! To help you, I've listed some good references at the end of the chapter. For students who feel less confident in math, there are also reviews of foundational topics that would not normally be covered in introductory mathematics courses at university, such as a recap of fractions and basic algebra skills. Readers who are familiar with these concepts can of course skip this material. The review and discussion of algebra is followed by a short introduction to probability and statistics, an overview of how to make good graphs, and then some thoughts on how you can estimate what a particular quantity might be.

5.2 MATH TAKES PRACTICE

Math is something we become better at by doing it repeatedly. Think about it this way: if you want to be a concert pianist, you can't just go to lots of concerts hoping to develop the skill just by listening. We all know that mastering the piano requires that you practise, practise, practise!

Math is just the same. For some it comes easier than others, but pretty much everyone with the grades to enter university science is capable of developing the math skills needed for an undergraduate science degree.

☠ DANGER ZONE

It is nearly impossible to "cram" (meaning to leave all your studying until the last minute) for a math or science exam. Because of the nature of the sciences, and the fact that you need to practise applying science and math skills in order to learn them, students that leave their studying until the last minute usually end up doing poorly on tests and exams in math and sciences. So plan your studying time accordingly (more on that in Chapter 7).

It's really good to think of math as a tool kit. The more tools you have at your disposal, the better off you'll be when answering questions. That means that working on math that you don't understand very well is actually a really good idea—you'll have more tools you can use on any given problem. You probably found in high school that some concepts in math seemed easier than others, and university will be the same. When you work on things that seem really hard, it's easy to get frustrated and discouraged, but if you put the effort in and eventually "get it," you'll find that it really pays off. Just being really, really good at **trigonometry**, for example, isn't actually going to get you very far when you actually need **calculus** to solve a given problem.

If you've made it to university, you should be capable of learning the math needed for science, although I should mention that there is actually a rare diagnosed disability associated with mathematics called *dyscalculia*. It can affect people that are very high-functioning in other areas of reasoning. What's more, because **dyscalculia** has a number of different symptoms, it is quite possible for individuals to have problems with specific tasks, mental arithmetic being one example, and yet prove to be incredibly adept at other kinds of math.

❓ Did you know?

The brilliant mathematician David Hilbert (1862–1943) was notorious for having poor mental arithmetic skills, although whether he actually suffered from dyscalculia is not clear since the disability was first diagnosed in the 1970s.

5.3 MATH IN PHYSICS, CHEMISTRY, AND BIOLOGY

While everyone knows that physics relies heavily on formulas and algebra, both biology and chemistry also use mathematics extensively. The key difference comes from the fact that biological and chemical systems are frequently made of many components whose interactions cannot be described with simple mathematics. The acceleration of a body under a force is easily described with Newton's second law of motion, but the detailed physiological responses of a frog to exertion cannot be described by a single simple formula.

This all really comes down to how complex things are. A single atom is actually a pretty simple system. In the most common form of hydrogen, there is just one proton in the nucleus and one electron orbiting it. The physics of atoms is described by the theory of quantum mechanics, and we can solve all the equations for the hydrogen atom using well-known methods. To a lesser extent, we can also describe the interactions of compounds and atoms more complex than hydrogen using well-understood math, so chemistry is actually similar to physics in its use of math. But for biology, take the human body as an example—it contains tens of trillions of cells, each with specific roles, and each cell contains an equally large number of atoms. By any stretch that's a pretty complicated system—Dr. Richard Dawkins has even suggested that the human brain is the most complicated thing in the Universe. Thus, biology has to handle some really tough challenges—it's no surprise that most of the time we don't start biology by writing down equations.

These different areas of application also affect the type of mathematics used most often in the different fields. In general, but not always, physics tends to be described most accurately through methods derived from calculus. In some sense, calculus evolved from considering how one thing changes in relation to another, an idea that describes many physical processes. When you're using calculus, it's also necessary to be good at algebra and know how to solve equations. Anyone that chooses to specialize in theoretical physics has to be a really good mathematician.

Chemistry has similar math requirements to physics. There's slightly less focus on calculus in the first two years of a chemistry degree, but you will still spend a lot of time considering ratios and **unit conversions** as well as solving equations to determine what reactions are possible. A typical chemistry degree will require a number of math courses—including calculus.

Math requirements in biology tend to be somewhat different. While there are obviously fields of biology that use calculus and solve equations (population dynamics is a good example), the most commonly used mathematical tool in biology is undoubtedly statistics.

But it's important to remember that statistical analysis (e.g., Bevington & Robinson, 2003) is a common part of *all* experimental sciences. Uncovering trends and relationships, as well as verifying how accurate a scientific theory is, are just two of the important applications of statistics. While statistical analysis cannot ever demonstrate that something is true in an absolute sense, scientists use probabilities of things being true all the time. As I discussed in Chapter 2, science has to work with theories that are known to be accurate to a certain degree. Hundreds of years of progress show that this works well!

5.4 IMPORTANT BASIC MATH SKILLS AND SOME TRICKS

Fractions Review

Fractions use a ratio of two whole numbers to represent a third number. The lower number, the **denominator**, specifies how the fraction divides things up. So examples like 1/2, 1/3, 1/4 progressively lead to smaller divisions of something; 1/4 of an orange is smaller than 1/2 an orange, for example. The upper number, the **numerator**, specifies how many bits of the divider you have. For the fraction 3/4, the denominator specifies that you are looking at quarters of something, while the numerator says you have three of them. If the numerator is larger than the denominator, then the fraction represents a number larger than 1 and is called an *improper fraction*.

Every fraction can be represented as a decimal number, although in some cases the decimal is infinitely long. For example, 1/3 is represented as a decimal as 0.33333…, where the 3 repeats an infinite number of times (it is infinitely *recurring*). Converting fractions to decimals is fortunately easy, you can just do the division within the fraction on your calculator and it will give you the decimal answer. Going the other way is more tricky, though, as there are actually many decimal numbers that can't be represented as fractions (see Box 5.1).

BOX 5.1   Irrational Numbers

Every fraction can be represented as a decimal number, although some of them are infinitely long. But the reverse isn't true—you can't write every decimal number as a fraction. We call the decimals that can't be written as fractions **irrational numbers**. The term "irrational" here means that the number can't be a ratio. The square root of 2 and π are just two of the more famous examples. Perhaps the easiest proof that these numbers exist is for $\log_2 3$ (see Section 5.4 for definitions if you're unfamiliar with logs). If this number can be expressed as a fraction, m/n, where m and n are integers, then we have $\log_2 3 = m/n$. Taking 2 to the power of each side, then $3 = 2^{m/n}$, and hence $3^n = 2^m$. But the right-hand side must be an even number, since all powers of 2 are even, and the left-hand side must be an odd number, since all powers of 3 are odd numbers. This logical contradiction means that $\log_2 3$ can't be written as a fraction!

Multiplying and dividing fractions are the two easiest operations you can do with them. Remember that to multiply two fractions you simply multiply the numerators and the denominators together. Here's an example:

$$\frac{1}{2} \times \frac{3}{4} = \frac{1 \times 3}{2 \times 4} = \frac{3}{8}$$

To divide fractions, we use the relationship that dividing by a/b is the same as multiplying by b/a. If this isn't clear, think about what happens when you divide by four. You can think about writing four as $4/1$. But dividing by four is the same as taking one quarter of something, or in other words multiplying by $1/4$. So if you want to divide two fractions, let's say $1/2$ and $3/4$, then you calculate as follows:

$$\frac{1}{2} \div \frac{3}{4} = \frac{1}{2} \times \frac{4}{3} = \frac{1 \times 4}{2 \times 3} = \frac{4}{6} = \frac{2}{3}$$

In the last step we used the idea of **equivalent fractions**. This idea says that if you have a fraction, a/b, you can always multiply or divide the numerator and denominator by the same number and get a new fraction that still represents the same number as the old one. It works because, by making the denominator and numerator larger or smaller by the same amount, you cancel out the changes. You can see this idea simply in the algebra:

$$\frac{2}{3} = \frac{2}{3} \times 1 = \frac{2}{3} \times \frac{2}{2} = \frac{2 \times 2}{3 \times 2} = \frac{4}{6}$$

Multiplying or dividing both the denominator and the numerator by the same amount is equivalent to multiplying by 1.

Sometimes people get confused when dividing by fractions produces a bigger number than you started with. But as long as you aren't using an improper fraction, that is exactly what should happen. To see this more clearly, let's look at what happens when we rearrange a division by a fraction. As an example, if we set

$$4 \div \frac{1}{2} = a$$

we can multiply both sides by 1/2 to get

$$4 = a \times \frac{1}{2}$$

This rearrangement immediately shows that the answer, a, has to be bigger than 4. Clearly in this case the actual value of a is 8.

Adding and subtracting fractions is slightly more difficult than multiplying and dividing them. Recall that when adding or subtracting fractions, you need to convert both fractions in such a way that they have the same denominator, or a **common denominator**. That explanation can sometimes be a bit confusing. But it's really simple. To add fractions, you must always add fractions of similar types, i.e., you add a certain number of quarters to another number of quarters. If you want to add 1/2 to 1/4, then you need to convert 1/2 into 2/4 first. (1/2 and 2/4 are equivalent fractions, as we discussed above.) For adding 1/6 to 1/9, you can see that both 6 and 9 are divisible by 3 (we say they have a "common factor"). The common denominator will then be the lowest multiple of three that both 6 and 9 will divide exactly into, which you can quickly check is 18. So to add 1/6 to 1/9, we first convert 1/6 to 3/18 and 1/9 to 2/18.

What about if you want to add 1/3 to 1/5? The common denominator is 15 because there is no number that divides into both 3 and 5. Although it isn't always a good thing to do, if you can't figure out a simple common denominator, you can always use the product of the two denominators, like the above example.

Once you've found the common denominator, you've actually done most of the work. It's then a matter of adding the adjusted numerators together. You never add the values of the denominators, because that wouldn't make any sense. The denominators represent how you have divided things up; the numerators tell you how many pieces of that you have. So adding the numerators tells you how many bits you have in the end. Here's an example:

$$\frac{1}{2} + \frac{2}{6} = \frac{1 \times 3}{2 \times 3} + \frac{2}{6} = \frac{3}{6} + \frac{2}{6} = \frac{3+2}{6} = \frac{5}{6}$$

In this example, the common denominator is 6, because 2 will happily divide into 6 three times. So we multiply both top and bottom by 3 to adjust this to a common denominator. If you're looking carefully, you should also see that 2/6 is actually equivalent to 1/3. If I had written 1/3 instead of 2/6 in the original example, the common denominator would still have been 6:

$$\frac{1}{2} + \frac{1}{3} = \frac{1 \times 3}{2 \times 3} + \frac{1 \times 2}{3 \times 2} = \frac{3}{6} + \frac{2}{6} = \frac{3+2}{6} = \frac{5}{6}$$

Subtracting fractions proceeds in exactly the same fashion, except that instead of adding the numerators, you have to subtract them. Here's an example:

$$\frac{1}{2} - \frac{2}{6} = \frac{1 \times 3}{2 \times 3} - \frac{2}{6} = \frac{3}{6} - \frac{2}{6} = \frac{3-2}{6} = \frac{1}{6}$$

That completes our quick review of fractions. While we've covered a few ideas, perhaps the key step to remember is that when adding and subtracting fractions you must find the common denominator first.

Unit Conversion and Dimensional Analysis

If you are frustrated by working with things in different units, that's OK. Believe it or not, many practising scientists feel the same way! But the fact is, some problems are more easily analyzed in one set of units than in others. For example, if you want to analyze the solar system, then expressing lengths in terms of astronomical units (approximately the **mean** distance between the Earth and the Sun, or 149 598 000 000 m) makes the numbers quite easy to handle. The Earth's average orbit radius becomes 1, and even Pluto doesn't get farther than about 49 astronomical units! But you really wouldn't want to use these units to describe the average size of a hydrogen atom. That's better done using a much smaller unit, namely the picometre (which is the same as one trillionth of a metre, or 1×10^{-12} m). Since many scientific fields use their own system of specially developed units, unit conversion is something that happens a lot in science, so it's important to know how to do it. While today you can use Google or Wolfram Alpha to do a lot of conversions for you, it's still good to know what is actually going on!

TOOL KIT

Wolfram Alpha is a tremendously powerful tool that can take mathematical and conceptual questions written in English, such as "what is half of 10?," and give you an answer. It isn't always correct on more difficult or advanced questions, and some of the answers can be quite funny, but it is perhaps the beginning of a new generation of powerful general-purpose computational tools.

The simplest form of unit conversion is when we go from one type of prefix on the unit (say centi-, kilo-, or mega-) to a different one. These conversions really just involve finding the right power-of-ten conversion from one unit to the other. But it's easy to forget what the prefixes represent, so some common SI prefixes are in Box 5.2 for easy reference, and are more fully explained in chapter 6.

IMPORTANT BASIC MATH SKILLS AND SOME TRICKS **127**

BOX 5.2   The Système International d'Unités (SI) Prefixes

Metric Measures and Imperial Equivalents

Unit	Measure	Symbol	Imperial Equivalent
Linear Measure			
1 kilometre	$= 1000$ metres $= 10^3$ m	km	0.62137 mile
1 metre		m	39.37 inches
1 decimetre	$= 1/10$ metre $= 10^{-1}$ m	dm	3.937 inches
1 centimetre	$= 1/100$ metre $= 10^{-2}$ m	cm	0.3937 inch
1 millimetre	$= 1/1000$ metre $= 10^{-3}$ m	mm	Not used
1 micrometre (or micron)	$= 1/1\,000\,000$ metre $= 10^{-6}$ m	μm (or μ)	Not used
1 nanometre	$= 1/1\,000\,000\,000$ metre $= 10^{-9}$ m	nm	Not used
Measures of Capacity (for fluids and gases)			
1 litre		L	1.0567 U.S. liquid quarts
1 millilitre	$= 1/1000$ litre $=$ volume of 1 g of water at stp*	mL	
Measures of Volume			
1 cubic metre		m^3	
1 cubic decimetre	$= 1/1000$ cubic metre $= 1$ litre (L)	dm^3	
1 cubic centimetre	$= 1/1\,000\,000$ cubic metres $= 1$ millilitre	$cm^3 = mL$	
1 cubic millimetre	$= 1/100\,000\,000$ cubic metres	mm^3	
Measures of Mass			
1 kilogram	$= 1000$ grams	kg	2.2046 pounds
1 gram	$= 1/1000$ kilogram $= 10^{-3}$ kg	g	15.432 grains
1 milligram	$= 1/1\,000\,000$ kilogram $= 10^{-6}$ kg	mg	0.01 grain (about)
1 microgram	$= 1/1\,000\,000\,000$ kilogram $= 10^{-9}$ kg	μg (or mcg)	

stp = standard temperature and pressure

For unit conversions of this type, you always need to know how many of one type of unit there are in another. For example, suppose you want to convert from 5.76 m to centimetres. In this case, you know that there are 100 cm in 1 m, so you can write down the conversion factor:

$$\frac{100 \text{ cm}}{1 \text{ m}}$$

Then if you take 5.76 m and multiply by the above factor to convert your units, you'll find that

$$5.76 \text{ m} \times \frac{100 \text{ cm}}{1 \text{ m}} = 5.76 \times 100 \frac{\text{m}}{\text{m}} \times \text{cm} = 576 \text{ cm}$$

What's happened is the two units of metres have cancelled each other out, leaving the value in centimetres alone. When working out the conversion factor, you have to remember that the bottom units will be the ones you want to convert from, while the top units will be the ones you want to convert to.

Another useful trick for remembering how to set up the conversion factor is that you always have to have the same amount, whether distance, time, or mass, on the top and bottom of the fraction, just written in different units. So in the above example, 100 cm for the numerator and 1 m for the denominator are the same physical distance. When converting between different unit prefixes, the conversion factor will always be a power of 10.

Probably the simplest mistake you can make is to get the unit conversion the wrong way around; for example, you might accidentally write 1 cm over 100 m. You can check for this kind of error by looking at your final answer. If you are going from big units to small, that is, from kilograms to milligrams, then you'll end up with a bigger value written in the smaller units. In this example, 1 kg is 1 000 000 mg. But if you go the other way, from small units to big units, then you expect the opposite. For our example, 1 mg is 0.000 001 kg.

Conversion between metric and non-metric units, or between units that are not expressed in powers of 10, is done in the same way, but it may involve a few more steps. In fact, as unit conversions become more complex, it is probably best to put the conversion factors in by looking at what units need to cancel. For example, suppose you want to convert a velocity of

3.6 m h^{-1} to metres per second. You can do this in two steps, converting through minutes and then to seconds:

$$\frac{3.6\text{ m}}{1\text{ h}} \times \frac{1\text{ h}}{60\text{ min}} \times \frac{1\text{ min}}{60\text{ s}} = \frac{3.6}{3600}\ \frac{\text{m}}{\text{s}} = 0.001\text{ m s}^{-1}$$

or, if you are feeling more confident, you can just multiply 60 s in a minute by 60 min in an hour to get 3600 s in an hour:

$$\frac{3.6\text{ m}}{1\text{ h}} \times \frac{1\text{ h}}{3600\text{ s}} = \frac{3.6}{3600}\ \frac{\text{m}}{\text{s}} = 0.001\text{ m s}^{-1}$$

For more complex conversions, it is best not to skip too many steps, though, as it is all too easy to miss a multiplication or division. In Box 5.3 we work through an example that changes both the length and time units.

BOX 5.3   An Example of Unit Conversion Involving Two Different Units

In this example, we'll convert a velocity, which involves changing both the length and time units. Specifically, we'll convert 36.0 mi h^{-1} to metres per second. The first unit we need is the number of metres in a mile. Looking that up, we find it is 1609.3, to one decimal place. Hence, the conversion factor for the distance is

$$\frac{1609.3\text{ m}}{1\text{ mi}}$$

The conversion factor for the time is

$$\frac{1\text{ h}}{3600\text{ s}}$$

This is the same as was used in an earlier example. Writing the complete conversion out, we get

$$\frac{36.0\text{ mi}}{1\text{ h}} \times \frac{1\text{ h}}{3600\text{ s}} \times \frac{1609.3\text{ m}}{1\text{ mi}} = 36.0 \times \frac{1609.3}{3600}\ \frac{\text{m}}{\text{s}} = 16.1\text{ m s}^{-1}$$

Unit conversions are used frequently in chemistry, in particular with stoichiometry (the branch of chemistry that looks at the amount of reactants and products in reactions). For example, in order to determine the number of grams of hydrogen in 2.03 g of water, you need to work through the following conversion:

$$2.03 \text{ g } H_2O \times \frac{1 \text{ mol } H_2O}{18.02 \text{ g } H_2O} \times \frac{2 \text{ mol } H}{1 \text{ mol } H_2O} \times \frac{1.01 \text{ g } H}{1 \text{ mol } H} = 0.228 \text{ g } H.$$

Dimensional analysis is a little bit different from unit conversion. It's a tool that can help you understand what a particular equation represents and also to check whether an equation you derive is consistent with what it should be. It's a very important concept in physics and chemistry as it can help you figure out what variables an equation should have. It may seem a bit dry, but it's incredibly useful.

The key idea is that equations describing scientific theories are independent of the units used for the variables in the equation. That's why we can choose different units despite using the same equations (we don't have different theories for different units!).

In dimensional analysis, we don't care about the exact units because we usually just describe quantities as possessing "dimensions" of mass, written M; length, written L; and time, written T. As an example, a velocity, which is length divided by time, has dimensions of L/T or LT^{-1}.

As a slightly more complicated example, let's consider applying dimensional analysis to Newton's second law of motion:

$$F = ma$$

where F is force, m is mass, and a is acceleration. The variable m is just mass, so it is represented by M. Acceleration has dimensions of length, L, divided by time squared, that is, multiplied by $1/T^2$, so a has dimensions LT^{-2}. Multiplying them together, ma has dimensions MLT^{-2}.

How does this compare to the left-hand side? Although this equation actually defines what a force is, we can check if its units are dimensionally correct compared to the right-hand side.

Since forces are often described using the newton SI unit, which is equivalent to kg m s^{-2}, we can immediately see that the dimensions of these force units are MLT^{-2}. Thus, both sides are completely consistent dimensionally, as they should be!

In Box 5.4 I give an example of using this approach to determine which equation describes the theory of gravity. This approach is actually used regularly to determine what the key factors might be in a physical theory, and, most important, it helps determine the particular number of factors that might be involved.

BOX 5.4   Example of Using Dimensional Analysis to Determine an Equation

Newton's law of universal gravitation describes gravitational forces as a function of the masses involved, their separation, and the gravitational constant, G. Given two formulas, $F = GMm/r^3$ and $F = GMm/r^2$, where F is the force, M and m two masses, and r the separation between the objects, you could determine which describes the gravitational force using dimensional analysis. You've seen that forces have dimensions of MLT^{-2}. So the right-hand side of the equation must have the same dimensions. The dimensions of G are M^{-1}L^{3}T^{-2}, and Mm has dimensions of M^2, so the dimensions of the numerator are ML^{3}T^{-2}. Depending on which formula you want to try, you would divide by L^3 or L^2. Then, replacing the variables with their dimensions for $F = GMm/r^3$ and $F = GMm/r^2$, and comparing them to the dimensions for force, you would find

$$\text{MLT}^{-2} \neq \frac{\text{ML}^3\text{T}^{-2}}{\text{L}^3} \quad \text{and} \quad \text{MLT}^{-2} = \frac{\text{ML}^3\text{T}^{-2}}{\text{L}^2}$$

So you have just shown that $F = GMm/r^2$ is the correct form.

Algebra Tips and Tricks

SOLVING FOR A VARIABLE AND EVALUATING A FORMULA. One thing that you will often have to do in science is rearrange a formula to express one variable in terms of others. If you have the values of all the other variables, then this is equivalent to solving for a particular value. There are essentially four tactics (there is actually one more I'll discuss later) that can be relied upon. Assuming $a = b$ as our starting point, we can

1. add the same amount to both sides of an equality: $a + c = b + c$

2. subtract the same amount from both sides of an equality: $a - c = b - c$

3. multiply both sides of an equality by the same amount: $a \times c = b \times c$

4. divide both sides of an equality by the same non-zero amount: $a/c = b/c$

Apply each of these possibilities step by step until you have rearranged the formula to give you what you want. As an example, Box 5.5 shows how a formula with six variables uses each of the steps outlined above.

BOX 5.5   Example of Rearranging a Formula to Solve for a Particular Variable

Rearrange

$$\frac{ab}{c} - d + e = f$$

to give *a* as a function of all of the other variables (this is sometimes referred to as isolating *a*).

The first step is to add *d* to both sides:

$$\frac{ab}{c} - d + e + d = f + d$$

The second step is to subtract *e* from both sides:

$$\frac{ab}{c} + e - e = f + d - e$$

The third step is to multiply both sides by *c* (don't forget the brackets around the right-hand side!):

$$\frac{ab}{c} \times c = (f + d - e) \times c$$

Continued

And the final step is to divide both sides by b:

$$\frac{ab}{b} = \frac{(f + d - e) \times c}{b}$$

giving the final answer

$$a = \frac{(f + d - e) \times c}{b}$$

To evaluate the numerical value of a formula, you need to follow a specific order. There's actually a great mnemonic for the order to use: **BEDMAS** (see Box 5.6). When we meet operations of similar type grouped together, such as addition and subtraction or division and multiplication, evaluation is always done in the order they appear from left to right. In fact, this ordering has a bit of flexibility, provided you follow a couple of *very* important rules. First, treat any subtraction as an addition of a negative number; for example, treat -4 as $+ (-4)$. Second, treat any division as a multiplication by the reciprocal; for example treat $\div 4$ as $\times (1/4)$. Let's apply these ideas directly to $1 + 2 - 4 + 3 = 2$. Provided you rewrite this as $1 + 2 + (-4) + 3 = 2$, you can rearrange the order of the additions however you like, for example, $3 + (-4) + 2 + 1 = 2$. For a multiplication-division example, $1 \times 2 \div 4 \times 3 = 3/2$ can be rewritten as $3 \times (1/4) \times 2 \times 1 = 3/2$. Remember, as long as you divide or multiply before you add or subtract, you're good. For big formulas, it's always a good idea to write down the numbers you get from your calculator at intermediate steps. If you hit one wrong button on your calculator, you're toast!

BOX 5.6   How to Remember the Order of Evaluations in Formulas: BEDMAS

B: Brackets or parentheses first
E: Exponents
D: Division
M: Multiplication
A: Addition
S: Subtraction

Example: Given $y = 3(2x - 2)^2 + 2$, where $x = 2$, what is the value of y? Start by substituting for x, and when you substitute for any variable it's always good practice

Continued

to put it into its own pair of brackets. That gives $y = 3[2(2) - 2]^2 + 2$. Following BEDMAS, it's clear that we need to evaluate anything in the brackets first. We now have two sets of brackets, so we must evaluate the inner ones first. That means evaluating $2(2) = 4$ before we do anything else. This leads to $y = 3(4 - 2)^2 + 2$. Now evaluate the next set of brackets: $4 - 2 = 2$. The next step is to evaluate the exponent, namely the square: $(2)^2 = 4$. The next piece is the multiplication by 3: $3(4) = 12$. Finally, you add the two terms: $12 + 2 = 14$.

MULTIPLYING BRACKETS IN FORMULAS. Suppose you are given the expression $(x + 3)(y - 5)$ and asked to multiply out the brackets. In this situation, each term in the first bracket multiplies each term in the second bracket. While you can do it step by step, writing out terms, one really good way is to use a table:

	y	-5
x	xy	$-5x$
$+3$	$+3y$	-15

Summing the terms up, we find $(x + 3)(y - 5) = xy - 5x + 3y - 15$. Note that it doesn't matter which axis you put the two brackets on: we could equally well have put $(x + 3)$ on the horizontal axis. It's also important to make sure that you have the correct plus and minus signs when you write out the terms.

This approach becomes even more powerful when the brackets contain more than two terms. In that case you can just increase the length of one of the table axes. As an example, consider $(x^2 + x + 3)(y + 5)$:

	x^2	$+ x$	$+ 3$
y	x^2y	$+ xy$	$+ 3y$
5	$5x^2$	$+ 5x$	$+ 15y$

Again, to complete the exercise you just add all the terms in the table. If you have three brackets to multiply out, then you can do two of the brackets first, sum up the terms, and then multiply by the remaining bracket.

Logarithms and Exponents

These two mathematical ideas are used extensively in all areas of physics, chemistry, and biology. In biology and chemistry, the pH scale of acidity and alkalinity is related to the logarithm of the concentration of hydrogen ions in solution, while in biology the range of sizes of genomes is so wide that they're best discussed in terms of the logarithm of the number of **base** pairs. **Logarithms** are also used extensively to determine relationships between different variables in an experiment.

Exponents are fairly straightforward to understand. They're derived from the concept of writing down the square and cube of a number, say 4, as 4^2 and 4^3. The upper number, either the 2 or the 3 in this case, is called the *exponent,* while we call 4 the *base.*

Exponents follow simple rules of addition and subtraction. You can see how the multiplication rule follows by writing out a simple example:

$$4^2 \times 4^3 = 4 \times 4 \times 4 \times 4 \times 4 = 4^5 = 4^{2+3}$$

In this case, you add the exponents: $2 + 3 = 5$. For division, on the other hand, you have to subtract the exponents, as is seen below:

$$4^3 \div 4^2 = 4 \times 4 \times 4 \div (4 \times 4) = 4^1 = 4^{3-2}$$

Exponents multiply if you take powers of them. For example, the square of 4^3 gives

$$(4^3)^2 = (4 \times 4 \times 4) \times (4 \times 4 \times 4) = 4^6 = 4^{3 \times 2}$$

Exponents are not limited to being positive whole numbers. They can be assigned any number. When the exponent is less than zero, say -2, it corresponds to taking the reciprocal of a number to the indicated exponent, for example, $4^{-2} = 1/4^2$. Exponents that are not integers correspond to taking the root of a number. The easiest way to see why this is true is to use the power rule:

$$(4^2)^{1/2} = 4^1$$

which shows that something to the power of 1/2 is equivalent to taking its square root. But this also applies to roots other than square and cube roots—these are special cases. You can see the same applies for any exponent b (i.e., writing out $(a^b)^{1/b} = a$ indicates that taking something to the power of $1/b$ is the inverse operation of taking something to the power of b). From these results, it then follows that

$$a^{b/c} = \sqrt[c]{a^b}$$

Logarithms are closely related to exponents. Thus far we have looked at taking numbers to a certain power, say a^b, where we are given both the base a and the exponent b. Let's allow this to equal some value c, such that $c = a^b$. But suppose we have a different situation where we are given c and a, but not b. In other words, what is the value of b such that when we raise the base a to the exponent b we get c? That's exactly the type of problem that logarithms solve for you.

You can get a feel for the definition of a logarithm by looking at how it relates to that of an exponent. Using $c = a^b$, logarithms are defined as follows:

$$\text{if } c = a^b, \text{ then } \log_a c = b$$

I've added arrows so you can see how the definition of a logarithm is like the inverse of taking an exponent. Let's apply this to a simple case, taking the logarithm to base 10 of 100. Since $100 = 10^2$, then

$$\log_{10} 100 = \log_{10} 10^2 = 2$$

which nicely shows how the logarithm function picks out the exponent. It's important to remember that although logarithms look like they only have one input number, namely the number that you take the log of, they also depend on the base.

Arguably the most common base used for logarithms is 10, which is written as either $\log_{10}$ or sometimes just log. This base is most common because our counting system uses base 10.

Another common base for logarithms is e, where $e = 2.718...$ is an infinitely long non-repeating decimal known as Euler's (pronounced *oilers*) number. Logarithms to base e are written $\log_e$ or just ln, for *natural logarithm*.

 Natural logs are important in a number of areas of science, from radioactive decay to chemical reaction rates to the growth of bacterial colonies (to name just a few examples). A number of short examples in the online material also focus on the correct approach to using calculators for these questions. Go to **science3.nelson.com**.

Because exponents obey some simple rules of addition and multiplication, logarithms do too. Since we know $a^b \times a^d = a^{b+d}$, then

$$\log_a(a^b a^d) = \log_a(a^{b+d}) = b + d$$

If we next substitute for b and d using $\log_a(a^b) = b$, $\log_a(a^d) = d$, then we find

$$\log_a(a^b a^d) = \log_a(a^b) + \log_a(a^d)$$

This is an important result for logarithms, namely that the log of a product of two numbers is the sum of the logs: $\log(AB) = \log(A) + \log(B)$. This result is true regardless of the base you use. A similar identity follows for the log of a quotient: $\log(A/B) = \log(A) - \log(B)$.

These two identities also lead to another useful identity for logarithms. If we assume that b is a whole number (to make things simpler, although it actually doesn't need to be), then using our above rule of summing terms that are multiplied with logarithms:

$$\log(a^b) = \log(\underbrace{aa \cdots a}_{b \text{ copies}}) = \underbrace{\log(a) + \log(a) + \cdots + \log(a)}_{b \text{ copies}} = b \log(a)$$

I have left off the base of the log in this case because the result is true regardless of the base you use.

The last cool thing that logarithms can do is change between bases. If you think about base 2 compared to base 4 logarithms, then since $4^b = (2^2)^b = 2^{2b}$, you can see that to convert a logarithm in base 4 to base 2, you multiply the exponent by 2. To understand what happens for a more general conversion, let's look at taking logarithms of the equation $a^b = c$, first using base a and then using base d. First, for base a we find

$$\log_a(c) = \log_a(a^b) = b$$

Then, using base d, we get

$$\log_d(c) = \log_d(a^b) = b\log_d(a)$$

and dividing both sides by $\log_d(a)$, we find that

$$b = \frac{\log_d(c)}{\log_d(a)}$$

Now we're almost there. Because we showed that $b = \log_a(c)$, we can use this result to get

$$\log_a(c) = \frac{\log_d(c)}{\log_d(a)}$$

This is the formula for converting the logarithm of a number c in base d into a logarithm in base a. To show how this works, consider our earlier example of base 4 and base 2 logarithms. In this case, the starting base, which is represented by d, is 4, while the base a we want to convert to is 2. The conversion factor is then given by the reciprocal of the denominator of our conversion formula: $1/\log_d(a) = 1/\log_4(2) = 1/0.5 = 2$. That exactly matches what we argued it should be earlier, and, importantly, remember this conversion formula is applicable for any base change, not just the example I chose.

To show a practical application of logarithms, in Box 5.7 I give an example of using logarithms in chemistry.

BOX 5.7 Example Calculation of pH Using Logarithms

pH is a tool used by chemists and biologists to describe the concentration of hydronium ions in a solution. Water molecules dissociate at a very low ratio of the total quantity: Of 10^7 molecules in solution, approximately 1 will split from H_2O to form an H^+ and an OH^- entity. This means a ratio of $1/10^7$ (or 10^{-7}) of molecules in a pure water solution will dissociate. The logarithm of this ratio is -7, and the convention for pH is to use the negative logarithm to represent acidity. If you have 10 times as many protons in the solution, making it more acid, you have 10^{-6} protons (note that this is a larger number of protons). The pH is therefore 6. If you have another 10 times as many protons, the concentration is 10^{-5}, which is pH 5. It's really important to note that the shift from pH 7 to 6 represents 10 times as many protons, while a shift from 7 to 5 is 100 times as many!

Matrices and Vectors

Because **vectors** and matrices are so incredibly useful at describing the world around us, I've provided a short introduction below. Depending upon your high school math curriculum, you may not have seen the material on matrices before, and you could see it in either first- or second-year math courses at university.

Why do we need vectors? Think about measuring the wind at any point on Earth. Not only does it have a very specific speed, it also has a direction (it could be north, northeast, or any other direction on a GPS!). The same idea applies to pretty much any form of movement—there is both a *magnitude* (i.e., size) and a *direction*—defining what we call a vector quantity. Quantities that have only a size are known as scalars. In a lot of cases we can solve problems without worrying about the direction part (you probably first met Newton's laws of motion this way), but that doesn't mean it isn't there!

Vectors have more than one component, but we can choose to represent them with one letter or symbol. There are actually a number of different conventions for writing vectors: They may be written in bold, e.g., **A**; with an arrow over them, e.g., $\vec{A}$; or even with an under-tilde, $\underset{\sim}{A}$. These symbols hide the fact that the vector has a number of components, but it can be written down easily. In Box 5.8, we work through a number of useful results for vectors.

BOX 5.8   Vector Conventions and Operations

Vectors can be written in a number of different ways in their component form. If $\mathbf{i}, \mathbf{j}, \mathbf{k}$ are unit vectors in the x, y, z directions, then we may write $\mathbf{A} = a_1\mathbf{i} + a_2\mathbf{j} + a_3\mathbf{k}$. This might also be written as a row vector, $\mathbf{A} = (a_1\ a_2\ a_3)$, or a column vector,

$$\mathbf{A} = \begin{pmatrix} a_1 \\ a_2 \\ a_3 \end{pmatrix}.$$

Vector magnitude: The magnitude of a vector is

$$|\mathbf{A}| = \sqrt{a_1^2 + a_2^2 + a_3^2}.$$

Vector addition: Each component adds to or subtracts from the other:

$$(1\ 4\ 5) + (3\ 2\ 1) = (4\ 6\ 6) \text{ in row format or } \begin{pmatrix} 1 \\ 4 \\ 5 \end{pmatrix} + \begin{pmatrix} 3 \\ 2 \\ 1 \end{pmatrix} = \begin{pmatrix} 4 \\ 6 \\ 6 \end{pmatrix} \text{ in column format.}$$

Vector multiplication: Vectors can be multiplied by a scalar:

$$3 \times (1\ 4\ 5) = (3 \times 1\ \ 3 \times 4\ \ 3 \times 5) = (3\ 12\ 15)$$

Dot product: The dot product takes two vectors and makes a scalar. Each component multiplies with the corresponding component in the other vector, and all products are summed at the end:

$$(1\ 4\ 5) \cdot (3\ 2\ 1) = 1 \times 3 + 4 \times 2 + 5 \times 1 = 16$$

Cross product: The cross product takes two vectors and makes a new vector. Each resulting component is actually a combination of 2 components from each vector. The general formula is

$$(a_1\ a_2\ a_3) \times (b_1\ b_2\ b_3) = (a_2 b_3 - a_3 b_2\quad a_3 b_1 - a_1 b_3\quad a_1 b_2 - a_2 b_1)$$

Vectors can also be multiplied by matrices to produce another vector. A **matrix** is an array of numbers $m \times n$ in size, with m rows and n columns. For vectors written in an x, y, z coordinate system the matrices are 3×3, and their entries are labelled as follows:

$$\mathbf{A} = \begin{pmatrix} a_{1,1} & a_{1,2} & a_{1,3} \\ a_{2,1} & a_{2,2} & a_{2,3} \\ a_{3,1} & a_{3,2} & a_{3,3} \end{pmatrix}$$

Row

Column

Notice that the indexing is actually the reverse of what you might expect if you are used to writing down coordinates as (x, y) pairs. The first subscript goes up as y goes down, while the second one goes up as the x position increases. But it's better not to think about these subscripts as being related to x and y; they're just numbers to label the position on a grid. The first number represents the row; the second number the column.

To multiply a matrix and a vector, we use the columns into rows rule—the column of the vector multiplies into the row of the matrix:

$$\begin{pmatrix} a_{1,1} & a_{1,2} & a_{1,3} \\ a_{2,1} & a_{2,2} & a_{2,3} \\ a_{3,1} & a_{3,2} & a_{3,3} \end{pmatrix} \begin{pmatrix} b_1 \\ b_2 \\ b_3 \end{pmatrix} = \begin{pmatrix} c_1 \\ c_2 \\ c_3 \end{pmatrix} = \begin{pmatrix} a_{1,1} \times b_1 + a_{1,2} \times b_2 + a_{1,3} \times b_3 \\ a_{2,1} \times b_1 + a_{2,2} \times b_2 + a_{2,3} \times b_3 \\ a_{3,1} \times b_1 + a_{3,2} \times b_2 + a_{3,3} \times b_3 \end{pmatrix}$$

This kind of operation transforms the input vector into another vector, and I give an example in Box 5.9.

BOX 5.9 **Multiplying a Vector by a Matrix: Example of a Reflection in the x-y Plane**

A reflection in the x-y plane can be represented in three dimensions by the following matrix:

$$\mathbf{A} = \begin{pmatrix} 1 & 0 & 0 \\ 0 & 1 & 0 \\ 0 & 0 & -1 \end{pmatrix}$$

Continued

If we take a vector $\mathbf{b} = \begin{pmatrix} 4 \\ 3 \\ 1 \end{pmatrix}$, then the multiplication $\mathbf{Ab}$ produces a new vector $\mathbf{c}$ that is the reflection of $\mathbf{b}$ in the x-y axis. Working through the calculation, we find

$$\mathbf{Ab} = \begin{pmatrix} 1 & 0 & 0 \\ 0 & 1 & 0 \\ 0 & 0 & -1 \end{pmatrix}\begin{pmatrix} 4 \\ 3 \\ 1 \end{pmatrix} = \begin{pmatrix} 1 \times 4 + 0 \times 3 + 0 \times 1 \\ 0 \times 4 + 1 \times 3 + 0 \times 1 \\ 0 \times 4 + 0 \times 3 - 1 \times 1 \end{pmatrix} = \begin{pmatrix} 4 \\ 3 \\ -1 \end{pmatrix} = \mathbf{c}$$

So the only thing that changed in the new vector is the z-coordinate.

Matrices can be used to represent many different kinds of transformations, including reflections, rotations, and stretches. Because you can think of doing one of these transformations after another, it should be no surprise that matrices can be multiplied as well. The rules for matrix multiplication are essentially the same as those for matrix–vector multiplication, except instead of one column vector, you now have three columns in a matrix:

$$\begin{pmatrix} a_{1,1} & a_{1,2} & a_{1,3} \\ a_{2,1} & a_{2,2} & a_{2,3} \\ a_{3,1} & a_{3,2} & a_{3,3} \end{pmatrix}\begin{pmatrix} b_{1,1} & b_{1,2} & b_{1,3} \\ b_{2,1} & b_{2,2} & b_{2,3} \\ b_{3,1} & b_{3,2} & b_{3,3} \end{pmatrix} = \begin{pmatrix} c_{1,1} & c_{1,2} & c_{1,3} \\ c_{2,1} & c_{2,2} & c_{2,3} \\ c_{3,1} & c_{3,2} & c_{3,3} \end{pmatrix} =$$

$$\begin{pmatrix} a_{1,1} \times b_{1,1} + a_{1,2} \times b_{2,1} + a_{1,3} \times b_{3,1} & a_{1,1} \times b_{1,2} + a_{1,2} \times b_{2,2} + a_{1,3} \times b_{3,2} & a_{1,1} \times b_{1,3} + a_{1,2} \times b_{2,3} + a_{1,3} \times b_{3,3} \\ a_{2,1} \times b_{1,1} + a_{2,2} \times b_{2,1} + a_{2,3} \times b_{3,1} & a_{2,1} \times b_{1,2} + a_{2,2} \times b_{2,2} + a_{2,3} \times b_{3,2} & a_{2,1} \times b_{1,3} + a_{2,2} \times b_{2,3} + a_{2,3} \times b_{3,3} \\ a_{3,1} \times b_{1,1} + a_{3,2} \times b_{2,1} + a_{3,3} \times b_{3,1} & a_{3,1} \times b_{1,2} + a_{3,2} \times b_{2,2} + a_{3,3} \times b_{3,2} & a_{3,1} \times b_{1,3} + a_{3,2} \times b_{2,3} + a_{3,3} \times b_{3,3} \end{pmatrix}$$

This might look like a complicated formula, but just by following the *columns into rows* rule, you'll see that it's actually pretty straightforward to write down. Sure, there are quite a few calculations to do to work it out, but it can be done by being systematic and not panicking.

IMPORTANT BASIC MATH SKILLS AND SOME TRICKS **143**

5.5 PROBABILITIES, STATISTICS, AND FITTING

Basic Ideas in Probability

If something is probable, it's likely to occur. But how can you quantify precisely how likely something is to occur? Guessing isn't helpful if you really want to know how likely you are to win a game of roulette in a casino! But the mathematics of probabilities comes to your rescue. Probabilities assign a number to how likely something is to happen, from 0, meaning something won't ever happen, to 1, meaning that it is certain. People also like to express the numbers as percentages, from 0 percent (won't happen) to 100 percent (will happen).

As long as something we want to study has a set of well-defined outcomes, such as heads or tails, or number values on a die, then we can work out the probability of each one of them happening. Suppose we roll a six-sided die that's numbered from 1 to 6 (so there are six possible outcomes when you roll the die). As long as the die is fair, then each of the numbers is as likely to be rolled as any of the others. If we focus for a second on rolling a 1, then the probability of it happening is given by

$$P(1) = \frac{\text{How many ways 1 can occur}}{\text{The total number of possible outcomes}} = \frac{1}{6}$$

where we use the notation $P(A)$ to represent the probability of a result A happening, and in this case we set A to be rolling a 1. If the die had two number 1s, then it would have two ways of coming up, but since it doesn't there is clearly only one way that a 1 can be rolled. Each of the numbers 1 to 6 is as likely to occur as the others. This means that the probability of rolling each of the other numbers is exactly the same as rolling a 1; that is, $P(1) = P(2) = P(3) = P(4) = P(5) = P(6)$. In general, the probability of a particular result occurring out of a number of possible outcomes is given by

$$P(\text{result}) = \frac{\text{How many ways the result can occur}}{\text{Total number of possible outcomes}}$$

☠ DANGER ZONE

You have to be precise when you describe a question in probability. Here's an example when the answers are quite different but the question looks similar:

1. If the probability of rolling a 1 is 1/8, and a fair die has 8 sides numbered 1 to 8, what is the probability of rolling each of the other numbers?
   Answer: Each of the other numbers must have the same probability as 1, namely 1/8.

2. If the probability of rolling a 1 is 1/8 and the fair die has 8 sides, what is the probability of rolling a number other than 1?
   Answer: The probability of rolling a 1 is 1/8, so the probability of getting any number other than 1 must be 7/8.

Being able to count the total number of possible outcomes is pretty important, and if you know that a set of results includes every possible outcome, then that set is called *collectively exhaustive*. If you can't evaluate how many possible outcomes there are, then you can't know how likely something is to happen. The formula we have been discussing also has one other important consequence: If you add up all the probabilities for all the possible results, the sum has to be 1. That makes sense: If you roll a die, you know you will get some kind of result! In addition, from the 0 to 1 scale it should be clear that if $P(B) > P(A)$, then B is more likely to occur than A.

❓ Did you know?

A good understanding of probability helps you to understand why most of the time spending your money on a big lottery ticket is really wasting it (unless you are doing it to support a charity, for example). Regardless of the number of tickets sold, assuming the lottery is by ticket number and not some choice of specific numbers, it's true that one of those tickets will win. But perhaps a better question is how many times, on average, would I need to enter the lottery to win it? Notice I've said on average here, as there is no certainty you will win regardless of the number of times you enter the lottery; it just gets more and more likely that you'll win as you enter more times. Suppose that a million tickets are sold and you buy one. Then the probability of winning is 1/1 000 000. (Just for comparison's sake, the odds of getting hit by lightning vary from 1 in 300 000 to 1 in 600 000, depending on which country you live in.) If you enter repeatedly, your probability of winning at least one time goes up. While the math is slightly tricky, the answer isn't: On average, you'd need to enter the lottery a million times to win it! Life is clearly too short!

Although we haven't discussed why, thus far we've implicitly assumed that probabilities add when looking at one experiment or trial. It's important to look at why this is, however. If we talk about the results from one roll of a die, we can talk about A happening or B happening, but clearly we can't talk about A and B happening at the same time. You can't throw both a 1 and a 2 at the same time with a die, so the events are said to be **mutually exclusive**. In this case, the logic of whether A or B happens is actually encapsulated in the law of adding mutually exclusive probabilities:

$$P(A \text{ or } B) = P(A) + P(B)$$

Coming back to the die again, the probability of throwing a 1 or a 2 is thus $1/6 + 1/6 = 1/3$. That should make sense to you: You are allowing for two results now, which means you should be twice as likely to get that particular outcome.

Combining Probabilities for Independent Trials

How do probabilities combine when we consider more than one trial? Suppose, for example, we toss a coin. What is the probability of throwing a head followed by another head? A key point here is that the two tosses of the coin are *independent* of one another. That is, the first result doesn't have any effect on the second result.

We know that the probability of throwing a head is $1/2$, so for this to happen twice in a row the answer must be the probability of it happening once multiplied by the probability of it happening again: $P(2 \text{ heads}) = P(\text{head}) \times P(\text{head}) = 1/2 \times 1/2 = 1/4$. You can even write it all out. If you toss a coin twice, there are four possible outcomes and only one of these has two heads: **HH,** HT, TH, TT.

So in general the rule of probabilities for independent events is

$$P(A \text{ and } B) = P(A) \times P(B)$$

Combining Non–mutually Exclusive Probabilities in a Single Trial

There are situations in a single trial where the outcomes A and B *can* both occur. Most often this happens when you consider questions about two or more properties of some-thing. Good examples are playing cards, which have both a number and a colour (and suit for that matter), or the distribution of student grades among males and females in the class. In this situation, we can classify the possibility of events occurring that include "and" logic; that is, what is the probability of picking a student who is male and received an A? Or what is the probability we pick a card that is red and a 4? These properties are clearly not mutually exclusive, and we write the probability for two non–mutually exclusive outcomes as $P(A \text{ and } B)$.

Given these ideas, we can think about asking questions such as, how likely it is that we'll pick a card that is red or a 4? Or how likely is it that we'll pick a student that has scored an A or is female (see Box 5.10)? In this case, you need to ensure that probabilities are not counted twice. The law of addition for non–mutually exclusive events is given by

$$P(A \text{ or } B) = P(A) + P(B) - P(A \text{ and } B)$$

BOX 5.10   Adding Probabilities for Results That Aren't Mutually Exclusive

Question: Suppose in a class there are 16 females and 14 males. In the distribution of grades, 14 students received A's, distributed to 8 females and 6 males. Suppose one student is then picked from the 30 in the class. What is the probability that this student is an A student or a female?

Answer: While it's tempting just to add the number of A students to the number of females in the class and then divide by the total number of students, that is actually wrong! You can see that because if you add the numbers up there are 16 females and 14 A students, which would suggest the probability is 1! But clearly you could pick a male student that didn't get an A from this distribution.

What's missing? It's actually a subtraction. You have to take into account that 8 of the females actually received A's to avoid including them twice (i.e., both in the count of the female students and also in the count of A students). These students are classified by the "and" logic—they both are female and received an A. The total probability of picking a student that is female or received an A is then given by

$$P(\text{female or } A) = P(\text{female}) + P(A) - P(\text{female and } A)$$

$$= 16/30 + 14/30 - 8/30 = 22/30 = 11/15$$

PROBABILITIES, STATISTICS, AND FITTING **147**

Conditional Probabilities for Dependent Events or Trials

In all the questions we have considered so far, we have ignored any sense of order, or more specifically *dependence,* in how things happen. But you may need to ask the question, "if A has happened, what is the probability that B will happen?" Such a probability is written $P(B|A)$, that is, the probability that B happens given A has happened. This is called **conditional probability**.

A really good example of conditional probability is picking marbles out of a bag and not replacing them. Once you take the first marble out, you've changed the number of marbles in the bag for any following selection. The same idea applies to picking cards out of a deck. Once you've taken one out, if you don't replace it, then the total number of cards has changed, which changes the probability of the second card being drawn.

In both cases we are talking about two trials, either picking one marble followed by another or one card after another. This means that the probabilities multiply, but now instead of simply having a probability for the second event B, $P(B)$, we now have the probability of B given that A has occurred, $P(B|A)$. The rule for combining dependent probabilities is thus

$$P(A \text{ and } B) = P(A) \times P(B|A)$$

which can be rearranged to give the law of conditional probabilities,

$$P(B|A) = \frac{P(A \text{ and } B)}{P(A)}$$

You will probably use a lot of probability theory if you study Mendelian genetics. The assortment of traits on chromosomes is typically analyzed in this way. For example, if both a man and a woman are carriers for cystic fibrosis, what is the probability that they will have a child with cystic fibrosis? To answer this question, you need to know that cystic fibrosis is an autosomal recessive disorder, so the fact that the parents are carriers means that each parent's genotype is *Aa* (*A* representing the normal allele, and *a* representing the cystic fibrosis allele). One allele from each parent gets passed on to the child. To have cystic fibrosis, the child needs to have the genotype *aa*. So the child has a 1/2 chance of getting *a* from Mom, and a 1/2 chance of getting *a* from Dad. $1/2 \times 1/2 = 1/4$. Thus, there is a 1/4 chance the child will have cystic fibrosis. You can also look at the four possible combinations to decide on the probability: *AA, Aa, aA, aa*. What I have just described is an example of non-conditional probability. Now, let's look at an example of conditional probability. Assume that you have a sibling with cystic fibrosis, but you are unaffected. What is the probability that you are a carrier for the CF mutation? If both of your parents are unaffected but are carriers (i.e., they are both Aa), and since the probability of being a carrier from their mating is 1/2, and the probability of being unaffected is 3/4, then you would calculate the probability of you being a carrier like this:

$$P(\text{carrier} \mid \text{unaffected}) = P(\text{carrier \& unaffected})/P(\text{unaffected})$$

$$P(\text{carrier} \mid \text{unaffected}) = (1/2)/(3/4) = 2/3$$

 Go to **science3nelson.com** for more examples of solving Mendelian genetics problems.

Introduction to Statistics and Fitting

Statistical analysis allows us to answer questions such as "how well do these data fit a theory?" or "how much variation can I expect in certain variables?" You're likely already familiar with some simple statistical ideas, such as an "average" representing the typical value you might expect for something. Fitting data is an important part of developing theories, as many times when given a data set we will want to describe relations between variables using an equation. One example is fitting a linear relationship, $y = mx + c$, to a set of data points.

Statistics and fitting are part of a very large field of study, so we can only give a short introduction here (but check the references for good resources). Even so, a number of key concepts, if you make them part of your methods tool kit, will make your experimental analysis stronger. Indeed, for clinical trials, statistical analysis is perhaps the second-most important step after experimental design.

VARIABLES, DISTRIBUTIONS, AND AVERAGES. In science, any quantity that you can measure or categorize is usually referred to as a variable. But variables can clearly come in different types. For example, they can have no logical ordering, such as the nationality of people in a room. Or they can have an order (say low, medium, or high) but no defined mathematical scale or spacing between them. Finally, variables can be mathematical values, for example, heights of people. Some people also like to make the distinction of categorical variables, namely **nominal** and **ordinal variables**, as compared to quantitative variables, namely **interval (or ratio) variables**. See Say What? for an explanation of these distinctions.

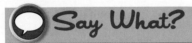
Say What?

NOMINAL VARIABLES: Variables classified by categories that have no defined ordering.

ORDINAL VARIABLES: Variables defined by categories that have an order or rank but for which no well-defined mathematical difference exists between the categories.

INTERVAL (OR RATIO) VARIABLES: Variables that are described by numbers where there exists a precise distinction (either an addition or a ratio) between possible variable values. These types of variables can also be subclassified into continuous, meaning all possible values are allowed, and discrete, meaning only certain numerical values (such as whole numbers) can occur.

When performing experiments, you need to understand the difference between the dependent variable and the independent variable (see Chapter 3 as well). The independent variable is the value that we have freedom to control in the experiment, whereas the dependent variable is a value that changes in response to changes in the independent variable. A simple example is adding mass to a hanging spring. Here the independent variable is the mass we put on the spring (we can add any mass that we want), while the dependent variable could be the measure of the extension of the spring.

Now that I've defined a number of important concepts surrounding variable definitions, we can think more precisely about how variables are distributed. There are perhaps two key, but very closely related, distribution concepts: *frequency distributions* and *probability distributions*.

The idea behind a frequency distribution, which often appears in table form, is very straightforward: Given a series of categories, how often does a given category appear in your experimental data? A simple example is the distribution of eye colour in a class of 24 students:

Eye Colour	Count
Brown	17
Blue	7

No doubt you've conducted similar numerical experiments in high school. From this frequency distribution, you can quickly work out the probability of a student having brown eyes if you select one randomly from the class. If the total number of students in the class is 24, then the probability of selecting a student with brown eyes is 17/24, and the probability of selecting a student with blue eyes is 7/24. We can then write out the probability distribution associated with the eye colour in table form:

Eye Colour	Probability
Brown	17/24
Blue	7/24

Thus, from a frequency distribution, we can convert to a probability distribution by dividing by the total number of data points taken (i.e., the total sample size). To denote a probability distribution, the following approach is often used. Suppose a variable under study can take a number of different values denoted by x, then the probability distribution $p(x)$ is given by

$$p(x) = \frac{\text{Frequency of } x}{\text{Total number of samples}} = \frac{f(x)}{n}$$

where we have defined $f(x)$ to be the frequency of x and n to be the total number of samples.

Over all the different possible values of x, the sum of all the $p(x)$ values must be 1. We write this using the Σ sign for a summation as

$$\sum_{\text{all } x} p(x) = 1$$

You may be thinking that the frequency and probability distributions are actually quite conceptually different—that's good! The **frequency distribution** is the result of an experiment, while the **probability distribution** tells us the probability of getting a result in a new experiment. But the great thing about probability distributions is that we don't just have to make them from experiments. We can infer how variables might be distributed. In fact many variables take on fairly similar distributions; for example, the heights and weights of people are distributed in statistically similar ways. From that assumed distribution we can calculate averages and work out how much variance might be expected.

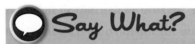 Say What?

FREQUENCY DISTRIBUTION: A table of results from an experiment showing the precise number of times a variable has been found to take a specific value.

PROBABILITY DISTRIBUTION: A function or table of values that assigns the probability of a variable taking a specific value.

At this point it is useful to review different kinds of averages. We talk about averages of things all the time, so you no doubt have a good idea of what one is—in a very approximate sense it measures the "middle" of some thing or things.

There are, however, a number of different averages. The one you are probably most familiar with is the arithmetic mean, the sum of the samples divided by the total number of samples. For a number of possible outcomes x_i, where i denotes the ith sample, over a total of n samples, the arithmetic mean is defined and written as

$$\bar{x} = \frac{1}{n} \sum_{i=1}^{n} x_i$$

But this can also be expressed in terms of the probability distribution $p(x)$ and the frequency that x occurs with, which we earlier wrote as $f(x)$. Look at the fact that the total number of samples, which is the sum of all the $f(x)$ values, divides the sum of all x values. That value can be moved inside the summation to give

$$\bar{x} = \frac{1}{n} \sum_{i=1}^{n} x_i = \sum_{i=1}^{n} \frac{x_i}{n}$$

This doesn't look like anything special. But actually it is the first step in a neat result.

The second step we need is to look at how the sum over all the events x_i relates to the sum over the possible values the x_i can take (which we wrote as x). Let's say the first three events all had the same value, for this example say 4, and the next three events took on the value 2. That means the sum of the first six events is $4 + 4 + 4 + 2 + 2 + 2 = 18$. But this sum is actually hiding in it the frequencies with which the values occur. There are three values of 4 (so $f(4) = 3$) and three values of 2 (so $f(2) = 3$). So this sum could equally easily be written $3 \times 4 + 3 \times 2$, which is just a sum over the values of x multiplied by the frequency of x. This means we can change a sum over x_i values to a sum over all the values of x as follows:

$$\sum_{i=1}^{n} x_i = \sum_{\text{values of } x} f(x) \times x$$

Let's now combine the two steps. Starting from the definition of the mean,

$$\frac{1}{n} \sum_{i=1}^{n} x_i = \sum_{i=1}^{n} \frac{x_i}{n} = \sum_{\text{values of } x} \frac{f(x)}{n} \times x = \sum_{\text{values of } x} p(x) \times x$$

where the last step, converting from $f(x)$ to $p(x)$, follows because of the definition of $p(x)$ I gave earlier.

To see all these rearrangements on some real data, I have created Table 5.1, which examines the results of rolling two dice.

The total at the bottom of the last column shows the average expected from the probability distribution, and you can confirm that it equals the average formula calculated by the frequencies by dividing the total of the third column, 694, by the total number of samples, 100. For a probability distribution, the arithmetic mean we have just shown here is given a specific name—the expectation.

Table 5.1 Outcomes from Rolling Two Dice

Sum of Dice Values (all possible x values)	Frequency of x, $f(x)$	Sum of These x Values, $x \times f(x)$	$P(X=x)$, $f(x)/n$	$x \times P(x)$
2	1	2	0.01	0.02
3	6	18	0.06	0.18
4	11	44	0.11	0.44
5	12	60	0.12	0.60
6	14	84	0.14	0.84
7	16	112	0.16	1.12
8	13	104	0.13	1.04
9	10	90	0.10	0.90
10	9	90	0.09	0.90
11	6	66	0.06	0.66
12	2	24	0.02	0.24
Total	$n = 100$	694	1.00	6.94

Although the **arithmetic mean** is discussed most often, there are other important averages, such as the median and mode. The **median** corresponds to the middle of a distribution of values of x. If you have a distribution of values 2, 3, 4, 5, 6, then the median is 4. It's also possible that there is no single middle number (if you have an even number of samples). The median is then the mean of the two middle values. For example, if your distribution of values is 2, 3, 4, 5, then the median is 3.5. The **mode** corresponds to the most common value in a distribution. For example, if your distribution of values is 2, 2, 3, 3, 3, 4, 5, then your mode is 3. Sometimes there can be more than one mode, as with 2, 3, 3, 3, 4, 4, 4, 5, or no mode, as with 1, 2, 3, 4, 5.

ARITHMETIC MEAN: Found by adding all the variable values and dividing by the total number of samples: $\bar{x} = \dfrac{1}{n} \sum\limits_{i=1}^{n} x_i$

MEDIAN: The middle value of the distribution, that is, the value that splits the distribution in half so there are as many values that are higher as lower.

MODE: The most commonly occurring value in the distribution, given by the value that has the greatest frequency.

UNDERSTANDING VARIATION. While averages, especially the mean, tell us much about what values a variable could be expected to take, they provide no information about how far from the average a given sample might be expected to fall. To understand this, we need a measure of how the variable varies or deviates from the mean. There are actually a number of ways to measure this, but the most commonly used method is the **variance** and its related quantity: the **standard deviation**. See Table 5.2 for calculating the variance of the two dice.

The key idea behind the variance is measuring, on average, how far from the mean the samples fall. If they fall far from the mean, then the variance is high; if they all fall close to the mean, the variance is low. We need to use a formula that will give us precisely this behaviour.

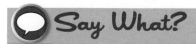
VARIANCE: Sum of the squares of the deviation from the mean of the distribution. Large values imply that the data are very spread out; small values that the data are clustered around the mean value.

STANDARD DEVIATION: The square root of the variance, and again a small value implies that the data are clustered around the mean, while a large value implies that they are very spread out.

Since the difference between the sample and the mean can be negative (i.e., the sample can be lower than the mean), we actually measure the average of the square of differences:

$$\text{Var}(x) = \frac{1}{n} \sum_{i=1}^{n} (x_i - \bar{x})^2$$

If we didn't square it, the values below the mean would offset those above and their sum would always tend toward zero—that isn't much use. Squaring means values both very much lower and very much higher than the mean make equally large contributions. If we have a probability distribution, then by exactly the same logic as shown for the mean, the variance can be expressed as

$$\text{Var}(x) = \sum_{\text{values of } x} p(x) \times (x - \bar{x})^2$$

Table 5.2 Calculating the Variance of the Two Dice

Sum of Dice Values (all possible x values)	Frequency of x, $f(x)$	Sum of the Squared Deviations, $f(x) \times (x - \bar{x})^2$	$P(X = x)$, $f(x)/n$	$p(x) \times (x - \bar{x})^2$
2	1	24.40	0.01	0.02
3	6	93.14	0.06	0.18
4	11	95.08	0.11	0.44
5	12	45.16	0.12	0.60
6	14	12.37	0.14	0.84
7	16	0.06	0.16	1.12
8	13	14.61	0.13	1.04
9	10	42.44	0.10	0.90
10	9	84.27	0.09	0.90
11	6	98.90	0.06	0.66
12	2	51.21	0.02	0.24
Total	$n = 100$	561.64	1.00	5.62

To get the variance from the total of column three, we divide by n, and find $\text{Var}(X) = 5.62$ (to three significant figures), in agreement with the sum of column five.

As well as the variance, statisticians like to use the standard deviation. The standard deviation is defined as the square root of the variance, and is often given the symbol σ. The square root is used because when we calculated the variance we had to use the square of the deviations rather than the deviations themselves. By taking the square root, we get a value that approximately measures the average actual deviation.

In what we've done so far, we've assumed that we are making measurements over everything we could measure. This means all the students in a class or, in other words, all of a given population. Very often we will not be able to measure all of the properties for the entire population, such as when we are polling people for how they will vote. A news agency can't possibly ask everyone in a given country a particular question. Instead it has to rely on a subset of people, which is called a sample of the population (frequently just called "a sample"—make sure you don't get confused over the different uses of "sample").

For each sample, we don't expect the measured variance to match that of the entire population. Neither, for that matter, do we expect the means to match. Indeed, different samples from the same population could have different variances and means. Clearly, in this situation both the mean and the variance could be expected to have their own kind of variance. Logically, we would want the average of all the possible sample variances to equal the population variance. As some moderately complicated math can show, it turns out that if we define the sample variance using $1/n$, then the average of all the sample variances does not actually equal the population variance. To make this happen, we actually need to define the sample variance using $1/(n-1)$, and the sample variance is often called s^2:

$$s^2 = \frac{1}{n-1} \sum_{i=1}^{n} (x_i - \bar{x})^2$$

It's important to note that the mean used here is the mean of the sample, which is defined using $1/n$, and not the population. If you take a large sample, then the difference between $1/n$ and $1/(n-1)$ becomes very small. There's a really good simple lesson here—the bigger your sample is, and provided it has been chosen well, the more likely you are to get numbers that are a good representation of the actual population.

GENERALIZING RESULTS FROM A SAMPLE TO A POPULATION: CONFIDENCE, SIGNIFICANCE, AND INFERENCE. By mentioning sampling from a larger population, we've touched on one of the most important aspects of statistical analysis—how likely is it that results from a sample are a good representation of the overall population? This is a classic example of what is termed *statistical inference*. This is a very big subject area, so we'll just look at some key issues that are important to know about.

One key point that we've not emphasized so far is that selecting a sample (a subset of the population) must be done in as random a fashion as possible.

TOOL KIT

For more information on how a sample of patients should be selected from a population, see Ben Goldacre's book *Bad Science* (Goldacre, 2010).

In the discussion of sampling a larger population, we pointed out that different samples would have different means and variances. But just how likely is your sample mean to be a good representation of the mean of the whole population? While we can't get the population mean from the sample, we can ask how likely it is that the sample mean lies within a certain range of the true population mean. Moreover, we can make our idea of "likely" more quantitative by setting what we call confidence. For example, is our sample mean expected to lie within a given range 95 percent of the time? The percentage value defines our confidence in the result (95 percent would mean it happens 19 times out of 20), and the precise range of values the sample mean falls in is what we call the *confidence interval*. The lower and upper limits of the interval are called the *confidence limits*. It's important to remember that the confidence interval is just something we're inferring. The population mean lies somewhere in that range; it isn't a defined parameter of the population in any way. Calculating confidence intervals relies upon making some very well-justified assumptions about how the variables will be distributed.

While confidence intervals are a great tool for understanding how likely a specific value is to be representative, sometimes we want to address much simpler questions, such as how likely it is that something is true or false. As it turns out, that can be a difficult question to answer for samples of data. This is because once we talk about sampling data, we have to factor in the impact of random chance in any of the values we find. For example, given two samples that show differences, how likely is it that those differences actually represent something truly significant as opposed to occurring just through chance? An excellent example is clinical trials: How probable is it that the control group results are different from those in the experimental group just by chance?

This is a really subtle and important point, so it's worth explaining in more detail. Usually in a trial the null hypothesis is that there is no difference between two groups (see Box 5.11). A question we can then answer is, "how likely is it that the hypothesis is true or false?," by testing how probable it is that the result could happen by chance alone. We don't have to work with just the null hypothesis either; you could assume the alternative hypothesis that there is a difference and then look at how likely it is that it is true.

BOX 5.11   Examples of Null and Alternative Hypotheses

Suppose we measure heights of people in samples from the north and from the south of the country. We try to ensure that the samples have as similar properties as we can (same age range and distribution of sexes for example). We then calculate the mean of the heights.

The null hypothesis is that any measured difference between the heights of the populations has arisen due to chance alone and that the populations have essentially the same mean height. You're inferring that there is no difference.

The alternative hypothesis is that there is a difference that has not arisen due to chance alone. In this case you are inferring that there is a real difference.

This is hypothesis testing, which we discussed in Chapter 3. Probabilities enter into things because we have no way of quantifying how a sample relates to an entire population. It's always possible when you randomly sample a group that through chance alone you produce a result that isn't representative of the overall population.

The "chance alone" wording is important here. If you wanted to know whether people from the east of the country were more likely to have brown eyes than those from the west, there always remains a possibility that your two separate samples might be quite different from the nature of the true populations. Although it seems unlikely, you might by chance pick 10 people from the west for your sample that all happen to have brown eyes, while the randomly selected people from the east could all have blue eyes. You're probably thinking that if we made the samples larger and larger those strange results would be less likely—good! That's true. But sometimes we can't do that either because of the cost or time or some other reason. So we have to be able to handle the uncertain nature of statistical sampling, and consequently any inferences we make have to be done in terms of probabilities. Inference doesn't provide absolute proof, but it does accurately quantify how likely something is to be true. And sometimes things are so likely to be true it would be foolish to suggest they aren't!

So how do we go about testing the hypothesis and evaluating probabilities? That's where the idea of significance testing comes in. Significance testing is slightly arbitrary for the following reason: When you discuss probabilities, at what point would you consider something to be very unlikely to happen? Would you think 1 in 4, that is, only 25 percent of the time? Or 1 in 10? Or even 1 in 100? There is no absolute answer here. So scientists usually take a value of 5 percent (i.e., 1 in 20) as representing something that is very unlikely to happen, but sometimes 1 percent is used instead.

This is the first step—deciding on the probability at which we think something is very unlikely to happen, which is known as the *significance level*. Importantly, you want to decide on this value before you start any experiment! You mustn't let any information from the experiment change your opinion of what the significance level should be. Five percent is the value commonly used, as stated previously.

Once you have made that decision, you want to know how probable it is that any difference you measure between the samples could occur by chance. In fact, significance testing considers how likely you are to measure a difference as big as, or larger than, what you found. This particular probability is called the *p-value*. There are a number of different ways of calculating it that rely on different probability distributions, but there's no room to discuss those details here. Instead let's focus on the decision-making process.

We now have two probability levels: the value at which we decide something is significant (our significance level) and the p-value, which tells us how likely a particular difference (or one that is even larger) is to occur by chance in our data. If that p-value is less than our significance value, then we say we have a *statistically significant* result. Our null hypothesis that things are the same appears very unlikely, so we reject it. If the p-value is greater than the significance level, then we can infer that the null hypothesis has not been rejected at our chosen significance level.

So how do these two possible choices relate to what is really true? Let's first consider the case that the null hypothesis is true, that is, that there is no difference between data sets. If you set a significance level of 5 percent, then 5 percent of the time you expect a p-value that appears statistically significant by chance alone, that is, your result is just a fluke of the sample you took. You would logically reject the null hypothesis in this situation, even though it was true. You can think of this as a false positive, and it is called a Type I error.

What if, on the other hand, the null hypothesis is false, that is, there is a difference, but you get a result suggesting that there is no difference? The probability of this happening depends entirely on the statistics of the two separate data sets. This is a Type II error. The types of errors are summarized in Box 5.12.

BOX 5.12   Rejecting or Accepting Hypotheses—Outcomes and Possible Errors

	Action	
	Null hypothesis accepted	**Null hypothesis rejected**
Null hypothesis is actually true	Good decision	Type I error: You concluded there is a difference when there isn't.
Null hypothesis is actually false	Type II error: You concluded there was no difference even though there is.	Good decision

I hope this analysis has made it clear why significance testing has to be treated with caution: You can never be absolutely sure whether a difference is real or a statistical fluke. But the good news is that tests can be repeated at higher significance levels using larger samples. The larger your sample is, the more likely it is that any random fluctuations will be ironed out.

FITTING. It is frequently important to be able to derive an equation that best represents a series of data. In science, this is a key step in verifying theories and the equations that represent them. Very often, the fitting is done within the context of experimental data, with accompanying error bars. While there are techniques for handling these error bars quite precisely, we'll focus on the main idea behind fitting in the context of a few data points.

In what follows we'll look at fitting the linear equation $y = mx + c$, since this type of fitting is straightforward and allows us to introduce some important concepts. The basis of the approach is that the best-fit line is the one that goes *as close as possible* to all the data points. It isn't about connecting the dots! But we need to decide what "as close as possible means" and for that matter how we can go about ensuring that.

The idea is actually very similar to how we construct a variance. Recall that we calculated a sum of squared deviations. It's reasonable to think about a best-fit line as minimizing deviations in the vertical direction from around the best-fit line (see Figure 5.1). This is the fundamental idea behind what is called a *least squares fit*.

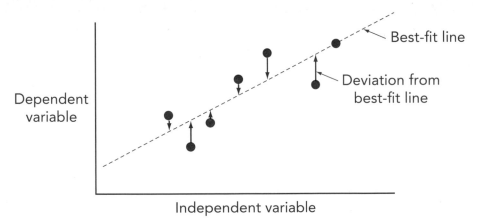

Figure 5.1   Descriptive graph of deviations in least squares fitting

Finding the exact line that minimizes the squared deviations is a problem that requires calculus, but it can be programmed on a computer very easily. That's good news too, as for a large number of data points the number of calculations required can become quite large.

While using a computer makes your life easier, it's important to remember that computers are dumb: The result you get out is only as good as your input and the questions you ask. So while you may be tempted to think that you don't need to know what's going on when fitting an equation, if you have an understanding of what's happening you'll be far more likely to catch mistakes in data entry and so on. And if you wonder where this might happen, remember that at some point you'll probably have to do a lab exam and make graphs under pressure!

 DANGER ZONE

Data entry errors are actually quite common, and very important to catch. In the United Kingdom in 2012, National Health Service statistics showed that 20 000 men had accessed obstetric care (Brennan et al., 2012). This absurd number was due to data entry errors. In the United Kingdom, the type of medical visit is entered as a three-digit code. So, errors could have been made when 560, indicating a midwife appointment, was entered instead of 460, indicating an eye doctor appointment. So, the lesson is to always double-check your data entry!

When line fitting is applied to statistical data, it is usually known as a linear regression analysis. A classic example, due to Francis Galton in 1885, examined the adult heights of sons or daughters relative to the average height of their mother and father, called the midparent height. Note that heights of daughters were also scaled up to account for women being shorter than men on average. As it turns out, tall parents tend to have children that are slightly closer to the mean midparent height, while shorter parents tend to produce children slightly taller than they are. This results in a best fit, or line of regression, that is shallower than $y = x$. The reason the word "regression" is used to describe this analysis comes from the fact that you could say the values "regress toward the mean." See Figure 5.2 for an example plot that uses data derived from Galton's original study.

PROBABILITIES, STATISTICS, AND FITTING **163**

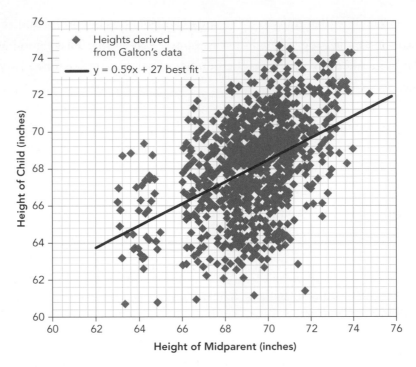

Figure 5.2 Regression toward the mean example using parent and child height data

5.6 MAKING GOOD GRAPHS

Imagine trying to explain experimental data with just a series of numbers. That would be pretty difficult and likely confusing. But take these data and display them on a graph, and you have something that can explain a host of things. From showing you relationships between variables to identifying where things might be going wrong, graphs are incredibly powerful.

But all of this assumes you've drawn or had the computer produce a good graph. This definitely doesn't happen just by typing in a few numbers or drawing a few lines on the page. You need to be careful and thoughtful when you draw a graph. The overall size of the graph on the page, the positions of the data points, your axis labels, and so on, tell the reader something about the information you are trying to communicate. If the graph is unclear, or at worst wrong, then you haven't achieved what you set out to do.

A good graph has three key components: a clear and descriptive caption, axis labels indicating both the variables and their units, and easily visible data points or lines. Additionally, if you are plotting experimental data, you will likely need to show error bars, while if you show more than one set of data, you should include a descriptive legend. It is traditional to plot the variable that you are controlling (the independent variable) on the x-axis and the dependent variable on the y-axis.

So what is a good caption? Well, "My data" isn't! Neither is "x versus y." You need to make it clear what data are being shown in the graph, so if you are plotting, for example, the maximum height a ball reaches relative to its starting kinetic energy, then clearly "maximum height" and "starting energy" need to be in the title. You might be tempted to suggest "Maximum height as a function of starting energy" as a good title. It's an improvement, but if there is a trend apparent in the data, and you've found that maximum height increases as a function of starting energy, then you should include that. So a really good title would be "Increase in maximum height as a function of starting energy."

The axis labels will also help the reader appreciate your plotted data more fully. In the above example, the x-axis should show the starting energy and the y-axis the maximum height. This is descriptive and helpful, but for the plotted points to be fully interpreted, you also need to specify the units used. If a value shows a maximum height of 3, is it 3 m, 3 km, or 3 ft? If the height was measured in metres, then the appropriate label is "Maximum Height (m)"; you may also see "Maximum Height/m." Be sure to find out which form is expected in your class.

You shouldn't make the plot too small. When drawing by hand, you don't have to use the whole page, but you do need to make sure you have enough grid points to fit the data in. The means looking at what your largest values are for both the x and y values. Give yourself enough room on the graph to include these points.

While it's important to be able to know how to put a graph together by hand, more and more today you will use electronic data capture and graph directly on the computer. Spreadsheet programs such as Microsoft Excel can produce many different types of plots for you. Your school may use a different approach to drawing graphs, and some programs are far more powerful than Excel. It's always good to experience different types of programs, as each has its strengths and weaknesses.

One last point about computer-generated graphs—make sure line widths are large enough that you can see them when you print them out. Sometimes lines can be so thin they are hard to see. The same goes for the font and size of the data points: Make sure they are clear and can be seen or read easily. See Figure 5.3 for an example of a good computer-generated plot compared to a bad one.

5.7 APPROACHES TO ESTIMATING

Very often in science you'll be faced with a question that is very difficult to get an exact answer for, but by a few careful deductions you can get a pretty good estimate. Some people, especially those without any kind of science training, may think that this amounts to "just guessing." But actually that's completely wrong. For a lot of things there are good limits on how big or small they can be. Similarly, we often know how something might be distributed. For example, coin tosses and dice rolling are random processes whose distributions we understand very well.

The key step in these problems is to determine how many pieces of information you need to estimate the number. Unfortunately, there is no general way of doing this, but again, experience with these types of problems is a really big help. Like all math tools, the more examples you see, the better you'll get at doing the problems. A really great introduction to this area is given in Dr. John Harte's book *Consider a Spherical Cow*. For now, let's start with a series of three simple problems, stepping up in complexity, to show how this kind of idea can work.

Suppose you have a swimming pool that needs to maintain a constant concentration of salt to provide the source for chlorination. Typically, this concentration will be about 3000 parts per million of salt relative to water.

Suppose the swimming pool has a volume of 150 m^3 and is filled with clean water to begin with. An interesting first question to ask is how many kilograms of salt do we need to add to the pool to get the desired concentration? Try to do this calculation and check your answer against Box 5.13 on page 168.

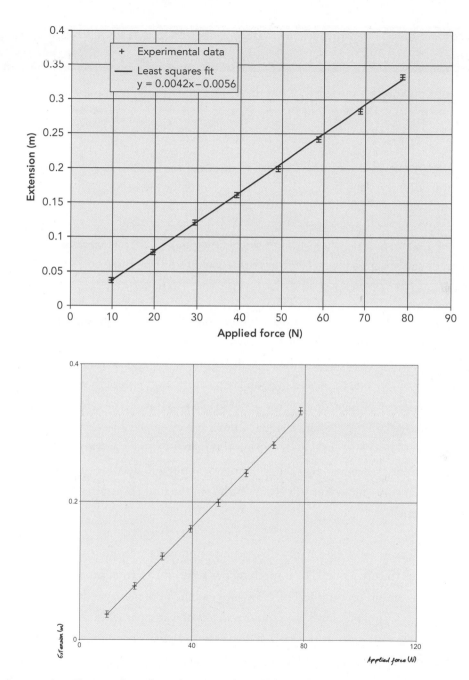

Figure 5.3 Example of a clear and well laid out graph (top) versus a poor one (bottom).

BOX 5.13 Calculating How Much Salt Is Needed to Reach a Given Concentration

Parts per million is a measure of the number of milligrams of something per litre of water (mg/L). The first thing to do is to work out the number of litres in the swimming pool—this is just a unit conversion from cubic metres (m^3) to litres.

$1 L = 10^{-3} m^3$, so the number of litres per cubic metre is $1/10^{-3} = 1000$. Thus, the total volume of the pool is 150 000 L. To calculate the amount of salt needed, we then take the 3000 mg L^{-1} and multiply by the volume of the pool to get 4.5×10^8 mg. Last, we need to convert this amount in milligrams to kilograms, another unit conversion, which amounts to dividing by 10^6. Hence, we arrive at 450 kg of salt. That's a lot of salt! But then again, it's a fairly big pool!

But once you've established the initial volume, water is lost through big splashes (we'll also say the pool is covered when it isn't used so evaporation isn't a big source of water loss). Let's assume a reasonable estimate is that each week a pool with moderate use will lose 2 percent of its volume. Assuming you keep filling up the pool to make up for these losses, how long, on average, does all the water in the pool take to be refreshed? This time period is what we call the average residence time. If we replace 2 percent per week, then we need to know how long it takes to replace 100 percent, which is $100/2 = 50$ weeks—almost a year. Note, of course, that the old and new water will become progressively more mixed so in fact some old water will always be left in the pool. What matters here is what the average time is.

In both these cases we've essentially divided one number by another to get the answer. That allowed us to turn a rate (i.e., something per time) into a total time period. But very often problems are more complicated. Now suppose the pool manager decides to put 1 kg salt per day into a pool that loses the same 2 percent of water per week. Assuming that the salt always mixes well within the pool, what will the eventual concentration of salt, after a few months of this procedure, be? That's clearly a harder problem, and we work through it in Box 5.14.

BOX 5.14 What's the Final Concentration of Salt if We Add 1 kg of Salt Each Day, While Losing 2 percent of the Water per Week?

While you might be tempted to think it depends on the initial concentration of salt in the pool, it doesn't! Over a long enough time, that initial concentration will be reduced or increased, depending on what value it started with, as the water and salt are replenished due to the losses.

To solve the problem, the first step is to determine how long it takes to refresh all the water in the pool, namely the residence time. But we've already done that—it's 50 weeks. This also has to be the residency time for the salt (because it is assumed to mix with the water). Thus, to determine the total amount of salt that mixes with the water in that time period, we need to calculate how much salt is added in $50 \times 7 = 350$ days. At 1 kg per day that works out to 350 kg.

The last step is to turn 350 kg in 150 m^3 into a parts per million (ppm) value. Since ppm is in milligrams per litre, the answer is 350×10^6 mg/150 000 L = 2333 ppm. That's actually pretty close to the required 3000 ppm value we mentioned earlier, so if you add a kilogram of salt a day to a 150 m^3 pool then you'll get fairly close to the required salt concentration over time.

Estimates like these can be very important in environmental science, because quite often the absolute numbers (such as how much pollutant was released into a river) are not exactly known. Working out concentrations like these is then an important part of a water management strategy.

But the application of estimation processes is far wider than this, and it is actually quite instructive to consider a somewhat different, but classic, problem. If there are a million people in a city, how many dentists would you expect to live in the city?

Let's assume the dentists that live in the city serve the people who actually live there! Little things like this can trip you up sometimes, but it's good to be aware of them. Then it's a question of working out how much work there is for dentists given a million people. If we go to the dentist on average once every six months, and spend 1/2 hour there, then each person requires about an hour of dental work a year. So we've estimated that the dentist needs 1 hour per person per year.

So we can now ask how many people a dentist can see in a year. Assuming that they work 7 hours a day for 5 days a week for 48 weeks of the year, that gives 1680 hours of work a year.

For our 1 hour per person per year that means each dentist can see 1680 people per year!

So if we divide 1 000 000 by 1680 we find that about 600 dentists are required for a city of a million people. How does this compare with actual data? Looking at the national average for Canada, we find there are 576 dentists per 1 000 000 people (Canadian Dental Association figures). That's very close to what we calculated!

Before you think this was a really good estimate, actually we've probably been a little bit lucky. There is quite a variation in these numbers, with some provinces having only 346 dentists per 1 000 000, and some reaching as high as 663 per 1 000 000. The variation within smaller towns and cities is probably quite a bit more. But as long as we take a large enough number of people, then the average works out pretty close.

It's important to remember when making estimates that there are limits to how much you should trust your figures. As a rule of thumb, the more things you have to multiply together, the more likely it is that the final answer could be a poor estimate. Take the process of doing the estimate seriously, but don't accept your answer as being the only possible one. It's always good to think about factors you might have missed. Check out *Consider a Spherical Cow* for some more great examples.

5.8 CONCLUSION

Whatever science you choose to do, sooner or later you will find yourself having to use mathematical tools. Whether it is the statistics of a clinical trial or calculus to determine the velocity of a rocket, math will help make a difficult investigation easier. While understanding the idea behind a scientific concept is very important, understanding the mathematics will very often give you a vastly more powerful tool for investigation.

The math presented in this chapter doesn't have any exercises to go with it; instead it focuses on trying to show how math is used in science. I hope you will spend time in math courses and get to study some of the topics discussed in greater detail.

FURTHER READING

Bevington, P., & Robinson, D. K. (2003). *Data Reduction and Error Analysis for the Physical Sciences* (3rd ed.). Boston: McGraw-Hill.

Brennan, L., Watson, M., Klaber, R., & Charles, T. (2012). The importance of knowing context of hospital episode statistics when reconfiguring the NHS. *British Medical Journal, 344,* 2432.

Goldacre, B. (2010). *Bad Science: Quacks, Hacks, and Big Pharma Flacks.* London, UK: Faber & Faber.

Harte, J. (1988). *Consider a Spherical Cow.* Los Altos, CA: University Science Books.

6

Physics, Chemistry, and Biology Basics

"And the actual achievements of biology are explanations in terms of mechanisms founded on physics and chemistry, which is not the same thing as explanations in terms of physics and chemistry."

MICHAEL POLANYI

6

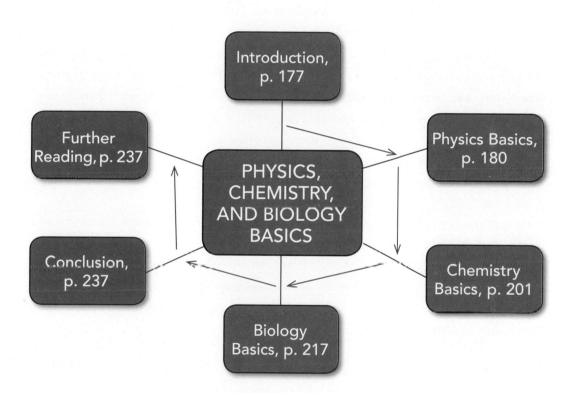

6.1 INTRODUCTION

The early chapters of this book focused on thinking like a scientist, understanding the language of science, and understanding the mathematical component of science. We've discussed at length that science is a process. However, in order to participate in that process, you need a core amount of background knowledge. The purpose of this chapter is to remind you of some of that knowledge, which you likely gained during your high school courses.

To write this chapter, I reviewed the high school science curricula in different provinces and states, and also examined first-year physics, chemistry, and biology courses at universities in Canada and the United States. This chapter contains a brief (with much emphasis on the word *brief*) highlight reel of the key concepts in physics, chemistry, and biology that I would expect you to know on entering my first-year class. Most first-year science courses at university require that you have had the corresponding high school science course. The high school course is thus considered a prerequisite, and you will build on the knowledge you gained in that course. Many of the topics you'll study in first-year university science courses will be the same as the topics you studied in high school, but there will likely be two differences: (1) University courses tend to go into more depth. (So, just because you've seen the topic before, don't feel overconfident with the material. You'll likely be taken deeper into the concept than you were before.) (2) University courses tend to stress application of knowledge in addition to accumulation of knowledge. Remember how we've talked about science being a process? Well, many university courses teach you about that process through asking you to apply the knowledge you've gained.

Note that this is not an exhaustive and comprehensive review of physics, chemistry, and biology key concepts. That would be impossible, especially given that many first-year science textbooks are more than a thousand pages long (that's for EACH of your physics, chemistry, and biology courses), and this chapter is only 66 pages long. This is just a highlight reel. If you see something in this chapter that is completely new to you, or that you don't understand, then be sure to review it before the start of class.

Throughout this chapter, we'll be discussing basic concepts of physics, chemistry, and biology. Each section will contain a brief history of the discipline, followed by a "Who's Who" section that profiles some key scientists of the discipline. Again, emphasis is on the word *brief*, as I'm trying to hit the highlights. It's impossible to list all the people involved in significant scientific advances of recent centuries, and no offence is intended if I've left your favourite scientist out of the Who's Who list. I'll then get into more details of the discipline by highlighting and briefly discussing the key themes of each discipline. At the end of each discipline is a list of common misconceptions that I see my students struggling with in class. As a professor, it is very important to me to identify and try to correct misconceptions early on in the course. Don't feel bad if you have any of these misconceptions. In fact, you should be happy that you've recognized them early on in your scientific education so that you can correct them and then build a foundation of new knowledge.

First, in Box 6.1, I explain SI (Système International) units, since they are common to the three disciplines of physics, chemistry, and biology. In 1875, some basic units of measurement were defined by international treaty. Since then they have been revised several times to provide greater accuracy of measurement.

BOX 6.1 Table of SI Base Units

Name	Symbol	Measure	Simplified Definitions
metre	m	length	The length of the path travelled by light in vacuum during a specific time interval.
kilogram	kg	mass	The mass of a prototype platinum-iridium bar in a vault in Paris, France.
second	s	time	The time required for the light emitted by a specific energy transition of a cesium-133 atom at rest to undergo 9 192 631 770 periods.
ampere	A	electric current	The constant current that under specifically defined circumstances produces a specific force per metre of length between two conductors.
kelvin	K	thermodynamic temperature	A specified fraction of the temperature at which water of a specified isotopic composition is present in the three states— liquid, gas, and solid.
mole	mol	amount of substance	The amount of substance of a system that contains as many elementary entities as there are unbound atoms, at rest and in their ground state, in 0.012 kg of carbon 12. You must always specify "moles of what," for example, moles of protons. Typically this number is taken to be Avogadro's number of items (6.022×10^{23}).
candela	cd	luminous intensity	The luminous intensity, in a given direction, of a source that emits monochromatic radiation of a specific frequency with a specific radiant intensity in that direction.

Let's continue with our overview of each discipline. We'll start with **physics**, and then proceed to chemistry, and then biology.

6.2 PHYSICS BASICS
Brief History of Physics

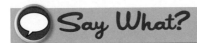

PHYSICS: The word *physics* comes from the Greek word *phusis*, meaning "nature." The "-ics" at the end of physics means "matters relevant to," from the Greek *-ikos*.

The study of physics as we know it today is a comparatively modern development that can be traced back to the Renaissance. If we go farther back in time it becomes difficult to disentangle physics (the study of the properties of space, time, energy, and matter) from the concept of natural philosophy, which is the study and analysis of nature in general.

It is possible to trace major developments in subfields of physics, such as astronomy and mechanics (the study of motion), back earlier. While we know Neolithic tribes built structures that were aligned with celestial events, the first detailed models of celestial motions were developed by the Greeks. Remarkably, despite incorrectly assuming Earth to be at the centre of the Solar System, Ptolemy's (CE 2nd century) model of planetary motions gave such good results it was used for hundreds of years afterwards. Given the intense growth of scientific concepts within Greek culture, spurred by a great interest in both rational argument and geometry, it is no surprise that the first theories of mechanics due to Aristotle (BCE 384–322) influenced the thinking on physics for thousands of years afterwards.

The ideas of the Greeks promoted advances in physics throughout civilization, from Asia to Europe. The next most influential events are arguably those that happened during the Renaissance. While Galileo revolutionized our knowledge of the heavens and steadfastly supported the heliocentric model for the solar system (i.e., Sun at the centre), Sir Isaac Newton contributed great advances in mechanics, gravitation, and optics. He also co-created one of the key tools of physics, namely the mathematical principles of calculus.

Not long after these developments, the Industrial Revolution and the growth of steam power made understanding the physics of these processes vital. This interest would lead to the development of the science of thermodynamics. Sadi Carnot contributed greatly to the understanding of heat engines. Later on, Ludwig Boltzmann would develop the theory of statistical mechanics, which provided a beautiful link between the microscopic behaviour of atoms and macroscopic thermodynamic properties.

Alongside these developments, interest in electromagnetism had been growing since the early 1700s. The experimental nature of this field meant that much was learned through occasionally quite dangerous experiments! While many individuals contributed important advances, the experimentalist Michael Faraday stands out. He is perhaps most famous for developing the first electric motor, but his discoveries and the equations he derived are even more remarkable given his comparative lack of mathematical education. His work inspired the elegant theory of electromagnetism developed by James Clerk Maxwell.

Toward the end of the 1800s there was a growing view that physics was almost in some sense "finished" and there were few discoveries to be made. In fact, nothing could be farther from the truth: The two pillars of modern physics, quantum theory and relativity, were waiting to be discovered.

? Did you know?

The British physicist William Thomson (later known as Lord Kelvin) gave a speech in 1900 in which he suggested that almost all the physical principles of interest had been discovered and there lurked only two unexplained phenomena, or "two clouds," as he called them. These two clouds turned out to be relativity and quantum mechanics. The speech is now infamous—he couldn't have been more wrong to suggest that physics was about to run out of new discoveries!

Over a period of 20 years from 1900 onwards, Albert Einstein would develop his special and general theories of relativity. While the special theory was an important step forward and reconciled a number of poorly understood concepts, the general theory provided a philosophical revolution. It turned humanity's understanding of the nature of space, time, and gravity on its head. Gravity was no longer a force, but rather a product of the geometry of the four-dimensional combination of space and time: space-time. The mathematical techniques used in the general theory were so advanced that few physicists at the time could understand it.

In exactly the same time period, the concepts of quantum mechanics began to take form. One of the key ideas of the theory is that physical properties, such as energy and angular momentum, cannot take on any value. Instead they are restricted to take on a specific set of values that often have a set increment (or *quantum*) between them. Max Planck provided the first theory to incorporate this idea, while Einstein, Niels Bohr, Erwin Schrödinger, and Werner Heisenberg contributed enormously to the theoretical, philosophical, and mathematical development of the theory. Quantum mechanics fundamentally challenged the traditional, or "classical," view of the world, as it was no longer possible to ascribe the properties of everyday objects, such as wavelike behaviour or particle-like behaviour, to the microscopic scale of atoms—small systems exhibit something known as wave-particle duality. For example, photons of light can appear to behave like individual particles in some experiments and like waves in others. A number of physicists did not take easily to this contradiction of their common sense!

The past 50 years have seen enormous growth in the applications of physics to technology and the study of the properties of materials (superconductivity being one example). At the same time, progress in understanding the fundamental interactions has seen the development of the four fundamental forces of nature, and even theories that combine the forces into so-called unified theories. Today, the most challenging problem facing theoretical physics is the difficulty of combining quantum theory and general relativity. While superstring theory remains a promising candidate, there is still much to be discovered in this area.

Who's Who of Physics

JOHN BARDEEN (1908–1991). A native of Wisconsin, Bardeen made seminal contributions to the applications of quantum mechanics to so-called semiconductor material, which resulted in the discovery of the transistor. He also helped pioneer theories of superconductivity.

NIELS BOHR (1885–1962). A Danish physicist who worked extensively on both the applications and philosophical implications of quantum mechanics, Bohr is as well known for contributions to nuclear fission as he is for co-developing the so-called Copenhagen Interpretation of quantum mechanics.

LUDWIG BOLTZMANN (1844–1906). Boltzmann was an Austrian physicist whose towering contributions to the study of statistical mechanics essentially gave birth to an entire field of study. Today, approaches he developed are applied to collections of atoms to predict the macroscopic behaviour of materials. His death is as sad as his contributions were great—he committed suicide at the age of 62.

SADI CARNOT (1796–1832). As the son of a military leader, his principal interest in physics came from a desire to improve French steam engine design. In a quite remarkable book that was largely ignored at the time, he laid out an abstract theoretical description of heat engines that paved the way for the development of the entire field of thermodynamics.

SUBRAMANYAN CHANDRASEKHAR (1910–1995). Many fields of astrophysics have been enriched by the contributions of Chandrasekhar. Growing up in India, he later settled in the United States and made contributions in fields from stellar structure through to hydrodynamics.

MARIE CURIE (1867–1934). Curie, a French-Polish chemist and physicist, extensively researched radioactivity, a term that she is credited with coining. In 1903 she received the Nobel Prize in Physics, which she shared with her husband, Pierre Curie, and Henri Becquerel for their work on radiation. In 1911, she earned a second Nobel Prize, this time in chemistry, for her discovery of and research on radium and polonium.

PAUL DIRAC (1902–1984). An Englishman who entered physics via engineering, Dirac is most famous for combining Einstein's theory of special relativity with quantum mechanics. Through this step he was able to accurately describe the behaviour of electrons as well as predict the existence of antimatter.

ALBERT EINSTEIN (1879–1955). A German by birth, Einstein later immigrated to the United States. His contributions to physics are extensive. From developing theories of relativity to helping found quantum mechanics, he is arguably without peer in terms of contributions to modern physics. Remarkably, his sole Nobel Prize is for the photoelectric effect rather than his theories of relativity.

MICHAEL FARADAY (1791–1867). An experimental physicist responsible for building the first electric motor, he also made important contributions to the mathematical description of electromagnetism and electrochemistry. Later in life he would become a professor of chemistry at the Royal Institution of Great Britain.

RICHARD FEYNMAN (1918–1988). A colourful individual who grew up in New York, Feynman is principally known for his contributions to quantum field theory via path integral methods. Hand in hand with this development, he introduced some unique approaches to evaluating the behaviour of particle interactions known as Feynman diagrams.

JEAN BAPTISTE JOSEPH FOURIER (1768–1830). A French mathematician and physicist, Fourier studied both vibrations and the transfer of heat and is widely credited with the discovery of the greenhouse effect. He is equally well known for developing an ingenious mathematical technique (Fourier analysis) that now forms an indispensable tool kit for studying the components of waves and signals, as well as being useful for solving differential equations.

GALILEO GALILEI (1564–1642). Galileo, an Italian astronomer, was the first person to point a telescope to the sky and make detailed observations of the Moon and the planets. He is famous for supporting the heliocentric (Sun-centred) model of the solar system to the point of conflict with the Catholic Church. He is sufficiently famous to be widely known only by his first name.

JOHANN CARL FRIEDRICH GAUSS (1777–1855). Gauss was born in Lower Saxony, Germany, and is perhaps more well known for his advances in mathematics than those in physics. Using techniques he developed for solving differential equations, he was able to predict the orbital positions of celestial bodies with great accuracy. He also made great contributions to electromagnetism—Gauss's law is the subject of many homework questions in electrostatics.

WERNER HEISENBERG (1901–1976). A German contributor to the foundations of quantum mechanics, Heisenberg developed a system of describing quantum-mechanical systems that is widely used today. He is perhaps most well known for his uncertainty rinciple: In quantum systems, knowing the position of a particle must inherently lead to an uncertainty in determining its momentum, and vice versa.

JAMES CLERK MAXWELL (1831–1879). Maxwell, a Scottish physicist, made seminal contributions to both electromagnetic theory and the kinetic theory of gases. Maxwell's equations are arguably one of the triumphs of nineteenth-century physics, containing relativistic ideas that predate Einstein's special theory of relativity.

SIR ISAAC NEWTON (1643–1727). Born a year after the death of Galileo in Lincolnshire, England, Newton is arguably the father of mathematical physics. As well as his ground-breaking laws of motion, he developed a theory of gravitation that continues to be used to this day, despite being superseded by Einstein's theory of general relativity. Equally important are his developments in mathematics, particularly calculus, which have had a profound effect on the evolution of science.

EMMY NOETHER (1882–1935). While she was arguably more a mathematician than a physicist, Noether's contributions to physics were nonetheless extremely important. Born in Bavaria, Germany, she developed Noether's theorem, which provides an explicit connection between symmetries of physical systems and conserved quantities.

MAX PLANCK (1858–1947). Planck was born in Kiel, northern Germany, to a well-educated family. He is most remembered for his contributions to the birth of quantum mechanics and his explanations of the so-called black body radiation, for which he received the Nobel Prize in 1918.

ANDREI SAKHAROV (1921–1989). A highly influential Russian physicist, Sakharov is principally credited with leading the design work of the largest thermonuclear weapons ever constructed by humans. Sakharov is also known for his later activism on civil liberties and reforms within the former Soviet Union, which ultimately led to his being exiled to the city of Gorky.

ERWIN SCHRÖDINGER (1887–1961). A phenomenal polymath of Austrian birth, Schrödinger contributed to botany and studies of Italian paintings. Today Schrödinger is known for the wave equation of quantum mechanics that bears his name. It describes how probabilities associated with different states of a quantum system change with time.

NIKOLA TESLA (1856–1943). While no single individual is responsible for the deployment of electricity in society, perhaps no individual has had a larger influence in this area than the Serbian Nikola Tesla. From the development of alternating current systems to futuristic ideas such as radio control of boats, his ideas were both revolutionary and visionary.

EDWARD WITTEN (1951–). Considered by many physicists to be the greatest living theoretical physicist, Witten has made numerous contributions across the fields of superstring theory and quantum field theory. He is also the first physicist to receive the Fields Medal for mathematics. He currently holds a position at the Institute for Advanced Study in Princeton, New Jersey.

HIDEKI YUKAWA (1907–1981). Arguably Japan's most famous physicist, Yukawa predicted the existence of mesons in a new theory of nuclear forces. Just a year later the predicted particles were indeed discovered, and a Nobel Prize for this research followed a few years later.

Common Physics Abbreviations and Notations

Although there are a large number of terms in physics, abbreviations tend to be quite specialized, and you won't meet that many in first-year physics. More important, each field tends to develop its own particular set of abbreviations that researchers are familiar with. However, abbreviations for units are standard. In Box 6.2, I have put together a number of common

BOX 6.2	Common and Not-So-Common Physics Abbreviations
2-D	two-dimensional
3-D	three-dimensional
A	ampere
AC	alternating current
BCS	Bardeen–Cooper–Schrieffer theory of superconductivity
BTU	British thermal unit

Continued

C	coulomb
C	heat capacity
c	speed of light, 299 792 458 m s^{-1}
CCD	charge-coupled device
CDM	cold dark matter
CFD	computational fluid dynamics
CMOS	complementary metal-oxide semiconductor
COG	centre of gravity
$\triangle d$	a small change in the variable d
DC	direct current
DE	differential equation
E	Young's modulus of elasticity
emf	electromotive force
EOS	equation of state
FEM	finite element method
FFT	fast Fourier transform
FTL	faster than light
FWHM	full width half maximum
Hz	hertz
IR	infrared
J	joule
K	kelvin
kg	kilogram
LMS	least mean square
LTE	local thermodynamic equilibrium
MRI	magnetic resonance imaging
MTBF	mean time between failures
N	newton
NIST	National Institute of Standards and Technology (U.S.)
NMR	nuclear magnetic resonance
P-P	peak to peak
Pa	pascal

Continued

PET	positron emission tomography
PDE	partial differential equation
QFT	quantum field theory
RF	radio-frequency
rms	root-mean-square
SI	Système International
SNR	supernova remnant
TEM	tunnelling electron microscopy
TNO	trans-Neptunian object
UHF	ultra high frequency
UV	ultraviolet
V	volt
W	watt
XRD	X-ray diffraction
ZPE	zero point energy

and not-so-common unit abbreviations that you might encounter. Note how many of the abbreviations in Box 6.2 overlap with abbreviations in the Common Chemistry and Biology Abbreviations boxes from later in the chapter.

? Did you know?

The word *laser* is an acronym for the phrase "light amplification by stimulated emission of radiation."

Box 6.3 will help you determine what a particular letter means in a given equation. Since physics often uses Greek symbols, we've also included some of the more common notations at the end of the table. But please remember that the definitions I give here are only a guide: Different textbooks might use a different convention. It's always important to double-check whether a book has its own specific definitions.

BOX 6.3   Common Physics Notation

A	area; amplitude
$\mathbf{a}$	acceleration (vector quantity)
C	capacitance; heat capacity; constant of integration
c	speed of light, 299 792 458 m s^{-1}; specific heat capacity
d	distance; diameter; differential
$\mathbf{D}$	electric displacement field (vector quantity)
E	energy
$\mathbf{E}$	electric field (vector quantity)
e	base of the natural logarithm, 2.71828…; electron charge, 1.602 176 565 × 10^{-19} C
$\mathbf{F}$	force (vector quantity)
f	frequency; friction; function
G	gravitational constant, 6.67384 × 10^{-11} m^3 kg^{-1}s^{-2}
g	acceleration due to gravity at the surface of the Earth, 9.81 m s^{-1}
h	height; Planck's constant, 6.626 069 57 × 10^{-34} J s
I	current; intensity; moment of inertia
K	kinetic energy
k	Boltzmann constant, 1.380 648 8 × 10^{-23} J K^{-1}
l	length
m	mass
$\mathbf{p}$	momentum (vector quantity)
q	charge
R	resistance
r	radius

Continued

S	entropy
s	speed
T	temperature
t	time
U	potential energy
$\mathbf{u}$	initial velocity (vector quantity)
V	voltage
$\mathbf{v}$	velocity (vector quantity)
w	width
α	fine structure constant, $7.297\ 352\ 569 \times 10^{-3}$
β	beta particle
γ	gamma ray; Lorentz factor in special relativity
δ	often used to represent a small change in something
ε	electromotive force
η	viscosity
θ	angle
λ	wavelength
μ	coefficient of friction
ν	frequency
π	pi, 3.141 59…
ρ	density
σ	standard deviation; dispersion
τ	mean lifetime
ψ	wave function in quantum mechanics
ω	angular velocity

Different Branches of Physics

It is worth noting that there are a couple of different ways of classifying physics that sit above the distinctions of the different branches. The first is the separation of theoretical physics, which relies on equations and mathematical models, from experimental physics, which takes an empirical approach to studying phenomena. In fact, these two methods are entirely complementary, and in many cases experimental and theoretical developments will work hand in hand.

? Did you know?

While many science subjects have traditionally been separated into theoretical and experimental domains, there is a growing belief that a third domain of investigation is emerging: computational science. Scientists working in this field simulate systems on computers and then use this information to develop better theories.

A second distinction, which is perhaps slightly more artificial, is between so-called pure physics and applied physics. The former is generally taken to mean the study of any physical phenomena with a view to developing new theories, while the latter is expressly reserved for applications of physical theories toward developing new technologies or devices. In reality, there is often significant overlap between the two fields, with new theories frequently being inspired by the development of new technologies.

I have divided the following list of the branches of physics by areas of application rather than by physics theories (Box 6.4). This means, for example, that quantum mechanics might be applied across a number of different branches, such as atomic physics, solid state physics, and astrophysics. Equally important, certain branches, such as astrophysics, rely on other branches, such as nuclear and atomic physics.

BOX 6.4 Different Branches of Physics

Acoustics: The physics of sound transmission in a variety of different materials.

Astronomy and astrophysics: One of the oldest branches of physics, astronomy is primarily concerned with making observations on celestial objects. Astrophysics applies theories of physics to explain these observations.

Continued

Atomic and molecular physics: The study and prediction of the properties of matter on the atomic and molecular level. This field often considers the interaction of light with matter and was one of the principal areas on which quantum mechanics had an immediate impact.

Biophysics: The application of theories and methods of physics to the workings and behaviour of biological systems. This field is one of the fastest growing in modern physics.

Chemical physics: Very closely related to atomic and molecular physics, but focused more on the quantum-mechanical explanation of chemical reactions.

Condensed-matter physics: The study of matter in a "condensed" phase, where the interactions between the constituent atoms are considered strong. This field has great relevance to technology because it often deals with properties measured at everyday temperatures and densities.

Electromagnetism: A theory that unifies the properties of particles with electrical charges or magnetic poles. This unification was one of the great achievements of nineteenth-century physics. Today this field contributes extensively to the design of communications technologies such as Wi-Fi.

Fluid dynamics: The study and theory of the motion of fluids (liquids and gases). Global climate modelling draws extensively on concepts from this field.

Geophysics: The study of the properties of Earth and its near environment. Plate tectonics (the explanation of how Earth's surface moves) is perhaps one of the greatest achievements in geophysics.

Mechanics: The theory and study of how bodies respond to forces. Mechanics has a very long history and today is studied primarily from a mathematical perspective.

Nuclear physics: The study of the properties and behaviour of the nuclei of atoms. This field explains why nuclear fusion and fission release such large amounts of energy.

Optics: The study of the behaviour and properties of light. It is one of the oldest areas of physics. Early theories relied upon raylike approaches with light travelling in straight lines, while later theories encompassed the more general wavelike properties of light.

Continued

Particle physics: Also sometimes known as high-energy physics. This field relies on quantum-mechanical theories to predict and measure the properties of subatomic particles such as the Higgs boson.

Plasma physics: The study of plasma, sometimes called the fourth state of matter, in which substances behave similarly to a fluid but contain electrically charged particles. The study of plasma is particularly relevant to the design of nuclear fusion reactors.

Solid state physics: The study of the behaviour of solid matter. While arguably a subfield of condensed-matter physics, solid state physics is particularly relevant to the creation of high-technology materials used in computer and communications technologies.

? Did you know?

The last Canadian winner of the Nobel Prize for Physics was Willard Boyle (1924–2011), in 2009, for the development of the charge-coupled device (CCD). This device for transferring electric current on a grid was developed into an imaging sensor and has greatly contributed to advances in imaging technology, particularly in astronomy.

Key Physics Themes/Concepts

KEY PHYSICS CONCEPT: CONSERVATION LAWS AND SYMMETRIES. This is a pretty high-concept idea to start with, but it is an incredibly powerful one that pervades almost all of physics. You've no doubt heard of the conservation of energy and the conservation of momentum. These two ideas, which you first saw in high school, allow you to develop theories of collisions, for example. The idea is powerful because you know you can equate momentum before and after a collision (for example) to figure out the velocities involved in the crash. Similar conservation laws apply in many fields of physics. For example, particle physics has many different types of conservation laws. Symmetries are a related idea in that they describe certain kinds of transformations you can make to a system without actually changing a particular property. For example, Newton's second law of motion surprisingly predicts identical behaviours for moving objects whether time runs forward or backwards! Very often symmetries like this also lead to a conservation law.

KEY PHYSICS CONCEPT: PARTICLES, WAVES, AND FIELDS FORM THE BASIS OF MECHANICS. To describe physical systems, physicists have to simplify systems under study by using a set of conceptual models. Waves, fields, and particles form the basis of descriptions that underlie the theory of mechanics: the study of how systems behave under the application of forces. Particles are simplified in the sense that we assume they have no spatial extent. Wave descriptions, such as vibrations on a string, can be conceptually more complex but encapsulate the position and time variation of the waveforms with comparatively few simplifications. Descriptions of fields are frequently even more complex than those for waves, but they can be applied to many situations. One particularly powerful example of a field description of nature is Maxwell's equations. This set of equations describes how electrical and magnetic fields evolve. It's worth noting that physicists debated whether light is a wave or a particle for hundreds of years until the discovery of quantum mechanics showed that it is both!

? Did you know?

The cosmic microwave background is a leftover field of radiation from the Big Bang, kind of a "cosmic echo" of that event. It contains information about the very earliest evolution of our Universe and is one of the most important research topics in modern astrophysics.

"Tune your television to any channel it doesn't receive and about 1 percent of the dancing static you see is accounted for by this ancient remnant of the Big Bang. The next time you complain that there is nothing on, remember that you can always watch the birth of the universe."

BILL BRYSON, *A Short History of Nearly Everything*

KEY PHYSICS CONCEPT: QUANTUM MECHANICS. The debate on the wave versus particle description of light raged from the 1600s to the beginning of the 1900s. Newton argued that light showed properties of particles, while his peers Hooke and Huygens demonstrated that many properties of light could be described by the properties of waves. At the turn of the twentieth century, a large number of physicists, but perhaps most importantly Albert Einstein and Max Planck, had the breakthrough that harmonized these competing ideas: the development of quantum mechanics. At its most simple level, quantum mechanics tells us that microscopic particles exhibit *both* wave- and particle-like behaviour and furthermore that properties such as energy and angular momentum can only change in discrete amounts or *quanta*. Despite having a significant number of philosophical challenges in its formulation, quantum mechanics has become one of the cornerstones of modern physics and describes the behaviour of microscopic systems with exceptional accuracy. From computational chemistry to the design and operation of computer components, quantum mechanics is behind an enormous number of advances in knowledge and technology and is arguably one of the two pillars of modern physics.

? Did you know?

Despite being one of the key contributors to the formation of quantum mechanics, Einstein objected to many of its philosophical implications. One particular part of the theory that essentially relies upon a random process led him to famously object, "As I have said so many times, God doesn't play dice with the world."

KEY PHYSICS THEME: THE FOUR FUNDAMENTAL FORCES OF NATURE. Decades of research have uncovered that our Universe appears to have only four distinct fundamental forces. In order from the weakest to the strongest, the forces are gravity, the weak nuclear force, electromagnetism, and the strong nuclear force. You are probably quite familiar with gravity and electromagnetism, but less so with the other two. The weak nuclear force is responsible for radioactivity in atoms and plays an important role in the nuclear fusion that powers stars. The strong nuclear force is the most powerful force in nature and holds together the nuclei of atoms. It has to be stronger than electromagnetism to overcome the electrical repulsion of positively charged protons. While these forces appear different at the matter densities and temperatures in our Universe today, the four forces are conjectured to merge together into one single force at incredibly high temperatures and densities. String theory is one current candidate for describing this force.

KEY PHYSICS CONCEPT: RELATIVITY. Einstein's theory of relativity is almost a buzzword in common culture. $E = mc^2$ is perhaps the most famous equation in all of science. Yet this equation is really just a tiny part of Einstein's elegant special theory of relativity and has comparatively little to say about his later general theory of relativity. At the root of the special theory are two key ideas: (1) that the speed of light is the same for all observers and (2) that observers who are in uniform motion relative to one another all have equally valid descriptions of physics. From these two ideas stems a remarkable series of consequences: Space and time should be thought of as similar parts of a larger four-dimensional space known as space-time. Lengths and time can change depending on your point of view. Relativity's predictions are as weird as they are amazing. The general theory extends these ideas to a far more complicated situation in which observers are allowed to have any kind of motion. Through adding more conceptual developments (most notably that someone falling in a gravitational field doesn't feel gravity), Einstein was able to develop a purely geometric theory of gravity. In the general theory of relativity, gravity is not a force but is instead caused by the changing shape (warping) of space-time due to the presence of masses such as planets and stars. The theory underlies our understanding of ideas from the evolution of the Universe to the nature of black holes. It has been tested in exquisite detail and shown to be accurate to the precision of our current experiments. If quantum mechanics is the first pillar of modern physics, then relativity should be seen as the second.

KEY PHYSICS CONCEPT: THERMODYNAMICS AND STATISTICAL MECHANICS. Thermodynamics is the science of heat flow and its relation to other forms of energy. This field plays an important role not only in physics but also in chemistry and biology. The field evolved primarily out of very practical concerns of developing a model to explain the efficiency of "heat engines," and the theoretical developments in this area begin with the seminal work of Carnot. Statistical mechanics allows us to relate these macroscopic properties to the collective behaviour of atoms on microscopic scales—a truly remarkable achievement. This connection was made by Boltzmann, who introduced probabilistic approaches to describing the behaviour of atoms. By combining the probabilistic approach with a sufficiently large number of atoms, Boltzmann was able to show that macroscopic properties could emerge naturally out of the mathematical descriptions he developed.

⚛ Key Physics Equations

I want to emphasize that merely remembering formulas is not a good strategy for passing physics courses. Physics is not "just math"; it's about knowing how a physical property can be represented mathematically. So you need to know the physical concepts *before* you relate them to the math. With that said, here are 10 important equations you'll likely need to use in first-year physics.

Newton's second law of motion, describing the relationship between force, F; mass, m; and acceleration, a:

$$F = ma$$

The definition of kinetic energy in terms of mass, m, and velocity, v:

$$KE = \frac{1}{2}mv^2$$

The potential energy for an object of mass m, at a height h, in a uniform gravitational field of acceleration g:

$$PE = mgh$$

Newton's law of universal gravitation, describing the force between two objects of mass m and M, separated by a distance r:

$$F = \frac{GMm}{r^2}$$

The centripetal force for a rotating system of radius r, mass m, and velocity v:

$$F = \frac{mv^2}{r}$$

The first law of thermodynamics, expressing the heat flowing into a system, ΔQ, as a function of the increase in the internal energy of the system, ΔU, and the work done on the system, W:

$$\Delta Q = \Delta U - W$$

The second law of thermodynamics, expressing the heat transferred in a system, ΔQ, as a function of its entropy change, ΔS, and temperature, T:

$$\Delta Q = T \Delta S$$

The velocity of a wave, v, as a function of its frequency, f, and wavelength, λ:

$$v = f\lambda$$

Ohm's law, relating the current, I, through a conductor to the resistance, R, and the applied voltage, V:

$$I = \frac{V}{R}$$

The energy of a photon of light, E, as a function of Planck's constant, h, and its frequency, f:

$$E = hf$$

Common Physics Misconceptions

MISCONCEPTION: THE NATURAL STATE OF OBJECTS IS TO BE AT REST. *Corrected Statement: Newton's first law of motion tells us that the natural state of motion is uniform velocity.* Explanation: This common misconception arises from the fact that friction on surfaces always tends to slow objects down. This idea influenced the Greeks heavily, and Aristotelian physics was built around a belief that heavy objects somehow want to be at rest. Newton's genius was to see through this misconception, and you could describe the thinking behind his first law as follows: In the absence of friction, objects want to keep on doing what they are doing.

MISCONCEPTION: MOTORS, WHETHER ELECTRIC OR COMBUSTION BASED, USE UP ENERGY TO CREATE MOTION. *Corrected Statement: The conservation of energy means that energy can be neither created nor destroyed. Motors convert one form of energy into another with varying levels of efficiency.* Explanation: The conservation of energy tells us that energy is never used up or lost. It can be converted into different forms, though, whether potential energy, kinetic energy, or different forms of heat. An electric motor converts electric energy into rotational motion but is not perfectly efficient. Some of both the electrical energy and the rotational energy is converted into heat through electrical heating of the wires and friction.

MISCONCEPTION: AIR IS WEIGHTLESS. *Corrected Statement: **The atoms in air have mass, and hence air has weight in a gravitational field.*** Explanation: Because the atmosphere extends kilometres above us, there is a common misconception that air is somehow weightless. While air exhibits the property of buoyancy, that is not the same as being weightless. Take a very strong box and pump all the air out of it and weigh it, and then perform the same test with air inside the box. You'll see that there is a difference. It's also important to emphasize that weight is a property that objects with mass have in a gravitational field. Mass exists regardless of whether an object is in a gravitational field.

MISCONCEPTION: WATER ROTATES IN DIFFERENT DIRECTIONS IN THE DIFFERENT HEMISPHERES OF EARTH DUE TO THE CORIOLIS FORCE. *Corrected Statement: **The Coriolis force has a much weaker impact on the rotation of the flow than either the initial water movement or the geometry of the bowl.*** Explanation: The Coriolis force is a real effect that is a result of Earth's rotation. It leads to hurricanes rotating in a counterclockwise direction in the Northern Hemisphere and in a clockwise direction in the Southern Hemisphere. But the strength of the Coriolis force is determined by the rotational speed of Earth—one rotation per day. This means that it is actually very weak compared to the rotation of water in a sink due to the flow from a tap, for example. Taps can make water rotate in a sink in a matter of seconds.

MISCONCEPTION: THE HOTTEST OBJECTS ARE RED AND THE COLDEST ONES ARE BLUE. *Corrected Statement: **The apparent visual colour of an object is related to the wavelength of its peak emission. For many objects this means that red is actually cooler than blue.*** Explanation: We naturally associate the colour blue with ice and therefore with cool temperatures. At the same time, we are used to seeing red flames and hence naturally associate red with something being hot. But the temperature of many objects—stars in particular—scales with the colours of the rainbow, moving from red (cool) to blue (very hot). Light at red wavelengths has lower energy and hence is related to lower temperatures than light at blue wavelengths.

MISCONCEPTION: ASTRONAUTS ORBITING EARTH ARE WEIGHTLESS BECAUSE THEY ARE SO FAR AWAY THEY FEEL NO GRAVITATIONAL FIELD. *Corrected Statement: Spacecraft orbiting Earth are constantly falling in Earth's gravitational field. Consequently, astronauts on the spacecraft feel weightless because when you're falling in a gravitational field you don't feel gravity.* Explanation: The acceleration due to gravity in lower Earth orbit is pretty much the same as that at Earth's surface. So astronauts don't feel weightless because they are somehow too far away from Earth. In fact they're falling all the time, but their spacecraft is moving so fast around Earth that the rate of descent is offset by the curvature of Earth's surface. If their motion around Earth is too slow, then they fall down to Earth's surface, while if their motion around is too fast they will go to a higher position or even escape Earth's gravity completely.

MISCONCEPTION: EXPLOSIONS IN SPACE MAKE ENORMOUS "BOOM" SOUNDS. *Corrected Statement: Sound cannot travel without a medium to pass through. Because there is no air in space, sound doesn't travel through space.* Explanation: Sound is a compression-type wave in a medium. It can travel at different speeds depending on what it is passing through. For example, the speed of sound in water is over 4 times that in air, while in iron it is about 15 times as fast. Without a medium (such as air) to travel through, that is, something to carry the compression, sound cannot travel anywhere.

? Did you know?

Stanley Kubrick's classic science fiction movie *2001: A Space Odyssey* is one of the few science fiction films where there is no sound portrayed in the vacuum of space. Even in the famous airlock scene, where an astronaut propels himself into the airlock before closing the outer door, no sounds are heard until the air rushes in to fill the airlock.

MISCONCEPTION: THERE MUST BE SOMETHING "OUTSIDE" THE UNIVERSE AND SOMETHING MUST HAVE HAPPENED "BEFORE" THE BIG BANG. *Corrected Statement: There isn't anything outside the Universe, and talking about events before the Big Bang may not make sense.* Explanation: We are used to understanding the positions of things by them being embedded in other things. For example, your house is on a street, the street is in a town, the town is in a province or territory, and so on. Extrapolating that mode of thinking leads us to think that the Universe must be within something larger. But we define the Universe to be the point at which that stops—there is nothing outside the Universe: The Universe is everything there is. This can all be done mathematically without any problems. In terms of what happened before the Big Bang, that's actually a very hard question. Time might be like the concept of pointing north. As long as you haven't reached the North Pole, you can always point in the direction you know to be north. But once you're at the North Pole you can't point north of it. Time could be similar.

 Go to **science3.nelson.com** for more examples of physics key concepts and downloadable tables and concept maps.

6.3 CHEMISTRY BASICS

Chemistry is usually defined as the study of matter and its transformations. Obviously, this also describes physics, and in reality there is a great deal of overlap between the two disciplines at the atomic level. I think the greatest difference is the point of view. While a key goal of chemistry is to use matter's internal structure (the arrangement and interrelationship of its parts) to describe the properties of matter, this generally delves into the subatomic levels only enough to help chemists understand the behaviour of atoms and how they form molecules. So a study of matter for most chemists includes understanding the arrangement of the energy levels of an electron in an atom, because without this it is not possible to understand the physical arrangement of atoms or molecules in space. Chemists also study the nucleus of the atom in nuclear chemistry, but the difference between nuclear chemistry and nuclear physics becomes almost undefined.

For a chemist, the structure of matter determines its properties. Properties of matter can be classified as macroscopic or microscopic and then as physical or chemical. Chemists observe matter to determine its properties and study the interactions of various substances with each other. They do this to understand why reactions take place, to determine how to control reactions, and to attempt to create new substances.

Chemistry is generally a quantitative science. Most relationships in chemistry can only be properly expressed in the language of mathematics. That is why high school and first-year university chemistry usually involves solving mathematical equations. Some areas of chemistry, such as organic chemistry, appear to use less mathematics but are still based on principles best expressed in the language of mathematics.

> *"It is a slightly arresting notion that if you were to pick yourself apart with tweezers, one atom at a time, you would produce a mound of fine atomic dust, none of which had ever been alive but all of which had once been you."*
>
> BILL BRYSON, *A Short History of Nearly Everything*

Brief History of Chemistry

Depending on your point of view, you could say that **chemistry** is as old as humankind. At some point someone observed changes in a piece of wood as it burned in a fire or noticed how mud hardened as it baked near a fire and then thought of ways that these reactions could be useful to them. By 4000 BCE, people had realized that heating certain bright blue stones produced copper metal. Between that time and 3000 BCE, someone thought of combining copper (a very soft metal) with tin to make bronze, with which it was possible to make weapons and armour. Metals were important in early civilizations, and the practice of metallurgy provided much chemical information.

But the chemical accomplishments of early civilization were not limited to metallurgy. By 3000 BCE, beautiful glass was being produced in Egypt, gunpowder was being used in China, and in India beautiful dyes were being produced for fabric. Perfumes were distilled in these early years using equipment that is still in use in some parts of the world.

In Western civilization, the beginnings of chemical theory emerged in about 600 BCE with the writings of Thales, a Greek philosopher. He proposed that chemical change was merely a change in the aspect of an element, his name for fundamental materials. Empedocles in 450 BCE was the first to write about the four elements: earth, air, fire and water. In fact, up until the seventeenth century, most scientists believed that all matter was composed of varying proportions of these four elements.

At the same time that Empedocles was writing about elements, other philosophers were asking if matter was continuous or discontinuous. Two Greek philosophers, Leucippus and Democritus, were the first we know of who advocated that matter was composed of small indivisible granules. In fact Democritus, in about 400 BCE, proposed the word *atomos* (*a-* means "not" and *tome* means "a cutting," so *atom* means "that which cannot be made smaller") for these granules of matter.

Say What?

CHEMISTRY: There are many possible theories for the origin of this word. Some believe that the word *chemistry* had its origin in China, being derived from the Hakka word *kim-mi* or the Cantonese word *kem-mai,* both of which have to do with searching for or making gold. Others believe it may derive from an old Persian word for gold, *kimiya* (600–300 BCE). Most people who worry about such things believe it originated with the Egyptian word for their own land, *Khem,* which meant black earth (3000–1800 BCE). The earliest use of the word *chemeia* or *khemeia* by the Greeks seems to have been in 400 CE. It is not known if this is an adaptation from the Egyptian *Khem* due to the mingling of Greek and Egyptian culture during the period 200 BCE to 300 CE or if it derives from the Greek word for casting metal, *cheo.* We do know that by 700 CE the word *al-kirniya* was being used in Arabic languages and that by 1150 CE the term *alchemy* was being used in Europe. In 1661, Robert Boyle coined the word *chymistry,* from which the modern word *chemistry* is directly derived.

From about 300 BCE to 1500 CE, the Greeks, Romans, Persians, Indians, Chinese, and Arabs were involved in the search for a process of transmutation that would convert one element into another. It was during this time that the art of **alchemy** came into being.

Say What?

ALCHEMY: Many popular books today explore some of the more flamboyant principles of alchemy. These include the search for how to make the philosopher's stone, which, according to practitioners of alchemy, possessed the power to turn lead in to gold and to produce the elixir of life, which would in turn give youth and immortality. What often gets missed in the discussion of alchemy is that many of the experimental techniques that were developed in this search for the impossible proved to be of great value and are still used today. Islamic alchemists also made important contributions to the development of the modern scientific method.

In Europe during the period from 1500 to 1650 CE, medical chemistry became much more significant to scientists than transmutation of elements. Diseases were rampant as cities grew larger and people lived in ever more crowded conditions. It is during this period that the seeds were sown for what is known today as the modern pharmaceutical industry.

By the mid to late seventeenth century, the behaviour and properties of gases were seizing the imagination of European scientists. An Irish chemist, Robert Boyle, discovered the relationship between the volume of a gas and its pressure, known by students today as *Boyle's law*. He eventually published a book called *The Sceptical Chymist* in 1661, which is full of experimental techniques carried out with a high degree of precision.

In 1789, A. L. Lavoisier published a textbook called *Elementary Treatise on Chemistry*. His incredibly precise and well-designed experiments led to what today is known as the law of conservation of mass. The experiments helped scientists throughout Europe to resolve uncertainties about elements, compounds, atoms, and chemical change, which had been sources of confusion and argument since the time of Aristotle (384 to 322 BCE).

By 1829, Johann W. Döbereiner, a German chemist, had laid the foundations for what would become the modern periodic table by establishing the idea of a systematic relationship among elements. In 1859, another German chemist, R. W. Bunsen (the inventor of the Bunsen burner), together with German physicist G. R. Kirchhoff, developed the spectroscope, which rapidly led to the discovery of many new elements. In 1864, English chemist John Newlands noted that if elements were listed in order of increasing atomic weight, there was a repetition of properties every eighth element. At approximately the same time, German chemist Lothar Meyer and Russian chemist Dmitri Mendeleev, working independently, discovered the idea of periodicity and published periodic tables of the elements. Meyer first published in 1864, but by 1869 he had a periodic table including 50 elements. Mendeleev first published in 1869 and by 1871 had published a version with elements listed in order of increasing atomic weight that left blanks in it for as-yet-undiscovered elements. Not only did he predict the existence of gallium and germanium, but he was, with his periodic table and the principles of periodicity, able to estimate very accurately what their physical properties would be. The modern periodic table is essentially the same as Mendeleev's table but with the elements listed in order of increasing atomic number rather than weight.

To begin to list the significant discoveries made in chemistry since 1871 would require a very large book of its own rather than just a portion of a chapter. Some of the big names in chemistry and little bit about what they did that was significant are given in the next section.

Who's Who of Chemistry

SVANTE ARRHENIUS (1859–1927). A Swedish chemist and physicist who is known for his theory of electrolytic dissociation, which led to the development of the Arrhenius equation, Arrhenius created a model of the greenhouse effect, established the field of electrochemistry, and was awarded the Nobel Prize in 1903.

AMEDEO AVOGADRO (1776–1856). An Italian physicist who was one of the first to distinguish molecules from atoms, Avogadro also studied the effect of combining volumes.

NEIL BARTLETT (1932–2008). An English chemist who was one of the first to make covalent compounds using noble gases while at the University of British Columbia in 1962, Bartlett later taught at the University of California, Berkeley.

JONS JAKOB BERZELIUS (1779–1848). A Swedish chemist who developed the symbols for many elements and calculated their accurate atomic weights, Berzelius also discovered selenium, silicon, and thorium.

JOHANNES BRØNSTED (1879–1947). Brønsted was a Danish chemist who, working independently from Lowry, introduced what became the Brønsted–Lowry definition of an acid as "something that donates a proton, and a base as something that accepts a proton."

STANISLAV CANNIZZARO (1826–1910). Cannizzaro, an Italian chemist, clarified the distinction between atoms and molecules and established the custom of using atomic weights in chemical formulas and calculations.

GEORGE WASHINGTON CARVER (1864–1943). An American botanist who is often listed as a chemist because of his work in devising products such as dyes, plastics, and gasoline made from peanuts, sweet potatoes, and soybeans, Carver started life as a slave but persisted in obtaining an education at a time when people of his race were rarely able to go to school.

He eventually obtained a master's degree from Iowa State University and spent much of his life helping others to obtain an education.

JOHN DALTON (1766–1844). Dalton, an English science teacher, is known for first proposing an atomic theory of matter and developing the laws of partial pressure.

HUMPHRY DAVY (1778–1829). An English chemist, Davy illustrated the connection between electrochemistry and the elements and discovered several elements, including potassium, sodium, barium, calcium, and magnesium.

PETER DEBYE (1884–1966). A Dutch physicist and chemist who did much of the early work in the field of polar molecules, Debye also made many other contributions to physics and chemistry.

THOMAS GRAHAM (1805–1869). Graham, a Scottish chemist, is best known for his work in dialysis, colloids, and the diffusion of gases, including Graham's law of diffusion.

FRITZ HABER (1868–1934). Haber, a German chemist, is widely known for the process he developed to make ammonia from nitrogen in the air. Despite an importation ban on nitrates to Germany following World War I, the Haber process allowed the Germans to continue making explosives. The process has also been used to make fertilizers, although the nitrate process is more common.

DOROTHY HODGKIN (1910–1994). Hodgkin, an English crystallographer, used X-ray crystallography to confirm the structures of penicillin and other biomolecules.

AUGUST KEKULÉ (1829–1896). A German organic chemist who is credited with describing the ring structure of benzene, Kekulé was one of the first chemists to realize how carbon atoms could link to each other to form complex structures.

IRVING LANGMUIR (1881–1957). An American chemist and physicist who invented a variety of things, including the incandescent lamp and a welding technique using hydrogen, Langmuir also helped Lewis's theory of atomic structure become more widely accepted and received a Nobel Prize for his work in surface chemistry.

HENRI LE CHATELIER (1850–1936). Le Chatelier, a French chemist, developed the principle named after him that states that a system in equilibrium will, when subjected to external forces, shift in the direction that will re-establish equilibrium.

GILBERT NEWTON LEWIS (1875–1946). Lewis, an American chemist, developed many bonding concepts, including the theory that covalent bonding is due to shared electron pairs.

THOMAS MARTIN LOWRY (1874–1936). Lowry, an English chemist, working independently from Brønsted, introduced what became the Brønsted–Lowry definition of an acid as "something that donates a proton and a base as something that accepts a proton."

WALTER NERNST (1864–1941). A German chemist and physicist who helped establish the field of physical chemistry and made major contributions to electrochemistry, thermodynamics, solid state chemistry, and photochemistry, Nernst is responsible for the electrochemical equation named after him.

ALFRED NOBEL (1833–1896). Nobel, a Swedish chemist, invented dynamite and used his personal fortune to posthumously found the Nobel Prizes, which recognize excellence in many fields of human endeavour.

LINUS PAULING (1901–1994). An American chemist who used quantum theory to determine the chemical structure of many compounds, particularly proteins, Pauling showed how electrons affect the formation of molecules and proposed the concept of electronegativity.

FREDERICK SANGER (1918–). An English biochemist who is the only person to have ever won two Nobel Prizes in chemistry, Sanger worked out the amino acid sequence for insulin and developed methods for elucidating the molecular structure of nucleic acids. He retired at age 65 and currently works on his garden in Cambridgeshire.

FREDERICK SODDY (1877–1956). An English chemist who worked with Ernest Rutherford on radioactive elements while they were both professors at McGill University in the early 1900s, Soddy also introduced the isotope theory of elements.

Common Chemistry Abbreviations, Notations, and Units

As in most other disciplines, there are many abbreviations and notations in chemistry. Some of them are so specific that it is best to wait until you study that particular area of chemistry before worrying about them. Others are fairly common and widely used; I will present these here (Boxes 6.5, 6.6, and 6.7). The difference between an abbreviation and a notation is that a notation is a symbol or letter that is used to refer to an idea or a specific word, whereas an abbreviation is just that, a shorthand method for writing a commonly used word or phrase. Please note that these tables are not meant to be exhaustive, although they may be exhausting! Note how many of the abbreviations in the boxes and table below overlap with abbreviations in the Common Physics Abbreviations box from earlier in the chapter, and the Common Biology Abbreviations box later in the chapter.

BOX 6.5   Common Chemistry Abbreviations

A	ampere
A	area
a	acceleration
C	coulomb
e	electron
F	Faraday
Hz	hertz
IR	infrared (referring to a region of the electromagnetic spectrum)
J	joule
N	newton
P	momentum; power; pressure
Pa	pascal
rf	radio-frequency
rms	root-mean-square
SI	Système International
UV	ultraviolet (referring to a region of the electromagnetic spectrum)
V	voltage; volume
V	volt
W	watt

BOX 6.6 Common Organic Chemistry Abbreviations

Ac	acetyl group
AcO	acetate
Bn	benzyl
Bu	butyl
DMSO	dimethylsulfoxide
EDTA	ethylenediamine tetra-acetic acid
Et_2O	diethyl ether
LAH	lithium aluminum hydride
Me	methyl
Ph	phenyl
n-Pr	n-propyl (the n is short for normal; normal-propyl or normal anything refers to a straight chain of carbon atoms as opposed to branched chains; it is less frequently used by chemists these days but still used in industry and other disciplines)
Pr	propyl
Py	pyridine
THF	tetrahydrofuran
THP	tetrahydropyranyl

BOX 6.7 Common Chemistry Notations*

[]	molar concentration
A	area; amplitude
a	acceleration
C	capacitance; heat capacity; constant of integration
c	speed of light; specific heat capacity; concentration
d	distance; diameter; differential

Continued

ρ or D	density
E	energy; electrical potential
e	base of the natural logarithm
F	force
f	frequency; friction; function
G	gravitational constant
g	acceleration due to gravity
h	height; Planck's constant
$H, \Delta H$	molar enthalpy (heat); a subscript is often used to denote heats of formation (f), heats of reaction (r), and other forms
I	current
J	coupling constant
KE or E_k	kinetic energy
K_{eq}	equilibrium constant; examples of constants with other subscripts are the acid ionization constant (K_a) and the solubility product (K_{sp})
l	length
m	mass
M	molar mass
mp	melting point
n	amount
PE or E_p	potential energy; photoelectric effect
Q	charge
R	gas constant
T	temperature
t	time

Continued

u	unified atomic mass unit, same value as amu
w	width
Δ	change in whatever symbol follows it; also used to denote heat when heat is added to a reaction

*Some symbols are used for more than one term, and some terms have more than one symbol in common usage. This is sometimes confusing for me, so I expect you may find it confusing too, but generally the context (or the equation you are using it in) will help you sort it out.

Different Branches of Chemistry

Just out of curiosity, I recently did a web search for listings of the various branches of chemistry. One site listed 50 different areas of chemistry. However, both the Chemical Society of Canada and the American Chemical Society list five main areas of chemistry that need to be taught in a university chemistry program in order for that program to be accredited. These branches are discussed in Box 6.8, with the inclusion of one other area rarely covered in an undergraduate program but increasingly important to chemists: theoretical chemistry.

BOX 6.8   Different Branches of Chemistry and a Brief Explanation of Each

1. **Organic chemistry:** The study of the properties and reactions of organic compounds, including all compounds with a carbon backbone.

2. **Analytical chemistry:** The analysis of chemical samples to identify their chemical composition and structure. Standardized methods to achieve this analysis are used in all areas of chemistry except theoretical chemistry.

3. **Physical chemistry:** The study of the physical and fundamental basis of chemical systems and processes. Important areas of study include thermodynamics, kinetics, quantum mechanics, statistical mechanics, and spectroscopy. Physical chemistry has a large overlap with molecular and chemical physics.

4. **Inorganic chemistry:** The study of the properties and reactions of inorganic compounds, which are traditionally those compounds not containing a carbon backbone. However, there is much overlap between organic and inorganic chemistry, especially in the area of organometallic compounds.

Continued

---

5. **Biochemistry:** The study of the chemical reactions and interactions that occur in living systems. Biochemistry is also associated with molecular biology and genetics.

6. **Theoretical chemistry:** The study of chemistry using fundamental theoretical reasoning. This involves a lot of mathematics, physics, and computational calculations.

🧪 Key Chemistry Themes/Concepts

KEY CHEMISTRY THEME: CHEMISTRY IS EVERYWHERE. Most, if not all, natural processes involve chemistry. Changes in the state, structure, and energy of matter maintain the world around us. At some level, in order to understand the world in which we live, we must understand chemical processes. For example, the chemical reactions determine the structure and function of cell organelles and ultimately the cell itself.

KEY CHEMISTRY CONCEPT: ALL MATTER IS MADE OF PARTICLES. One hundred eighteen elements have been identified, although several of these are human-made and can exist for only parts of a second. Of the 118, about 40 different atoms form the building blocks of the molecular and ionic structures that make up most of the known substances that chemists deal with.

KEY CHEMISTRY CONCEPT: THE PROPERTIES OF MATERIALS DERIVE FROM THE IDENTITY AND ARRANGEMENT OF PARTICLES. Atoms form or rearrange bonds during chemical reactions. The properties of the materials formed depend on which atoms are combined and the way they are arranged.

KEY CHEMISTRY THEME: ATOMIC THEORY. In order to understand chemistry, it is important to understand the structure of the atom. We know today that atoms are composed of a small, dense nucleus containing any protons and neutrons associated with the particular atom. Outside the nucleus is a "cloud" of electrons that move around the nucleus somewhat randomly. The energy associated with these electrons is described by mathematically derived energy levels. Much of the work done by chemists and physicists who study atoms is to understand how protons, which repel each other due to their like charges, are constrained to remain in the nucleus; the role of neutrons; and how to best describe or predict the behaviour of electrons. These forces are what we called the "strong forces" in the physics section earlier in this chapter.

KEY CHEMISTRY THEME: KINETIC MOLECULAR THEORY. The kinetic molecular theory of gases makes five simplifying assumptions about the behaviour of molecules in a gas. First, a gas consists of a collection of small particles travelling in a straight line that obey Newton's laws. Second, the molecules in a gas occupy no volume. Third, collisions between molecules are perfectly elastic (this means that no energy is gained or lost during the collision). Fourth, there are no attractive or repulsive forces between the molecules. Fifth, the average kinetic energy of all the molecules in the collection is proportional to the absolute temperature. From these simplifying assumptions arise the various gas laws, including the ideal gas equation.

KEY CHEMISTRY THEME: BONDING THEORIES. Various theories have been proposed over the past hundred years or more to explain how atoms form molecular or ionic compounds. Some of these are Lewis theory, electron domain theory, valence bond theory, and molecular orbital theory (MO theory). Of all of these, MO theory is the only one for which exceptions (or failures of the theory) have not been found. However, because this is sometimes a challenging theory to master, the other theories are still in use today to one extent or another.

KEY CHEMISTRY THEME: ENERGY IS CRUCIAL IN THE CHANGES THAT MATTER CAN UNDERGO. During physical or chemical changes, energy changes also occur as the bonds between atoms or molecules are re-formed. Catalysts are nature's way—increasingly mimicked by humans—of causing a reaction to occur with a lower energy expenditure and thus often at a faster rate.

> *"Chemistry without catalysis, would be a sword without a handle, a light without brilliance, a bell without sound."*
>
> ALWIN MITTASCH
> R. B. Desper (1948). Alwin Mittasch, *Journal of Chemical Education, 25,* 531–532.

Key Chemistry Equations

Arrhenius equation:

$$k = Ae^{\frac{Ea}{RT}}$$

Clausius–Clapeyron equation:

$$\log\left(\frac{P_1}{P_2}\right) = \frac{\Delta H_{vap}}{2.303R}\left(\frac{1}{T_2} - \frac{1}{T_1}\right)$$

Henderson–Hasselbach equation:

$$\text{pH} = \text{p}K_a + \log\frac{[\text{base}]}{[\text{acid}]}$$

Ideal gas equation:

$$PV = nRT$$

Nernst equation:

$$E = E^o - \frac{0.0592}{n}\log Q$$

Rydberg equation:

$$\frac{1}{\lambda} = R\left(\frac{1}{n_1^2} - \frac{1}{n_2^2}\right)$$

Van't Hoff equation for osmotic pressure:

$$\pi V_{solution} = n_{solute}RT$$

Key Chemical Reactions

In Box 6.9 I list some of the basic reactions in chemistry.

 ## Common Chemistry Misconceptions

In chemistry, as in all sciences, many ideas are easy to misinterpret. This is probably because so many of the concepts are very abstract. It is also true that ideas are constantly

BOX 6.9   Basic Reaction Types Typically Seen in General Chemistry

Type of Reaction	Description	Example
Synthesis or combination reaction	$A + B \rightarrow AB$	$2Na(s) + Cl_2(g) \rightarrow 2NaCl(s)$
Decomposition reaction	$AB \rightarrow A + B$	$2HgO(s) + heat \rightarrow 2Hg(l) + O_2(g)$
Substitution or single-displacement reaction	$AC + B \rightarrow A + BC$	$Zn(s) + 2HCl(aq) \rightarrow H_2(g) + ZnCl_2(aq)$
Double-displacement reaction	$AB + CD \rightarrow AC + BD$	$Cd(s) + 2HNO_3(aq) \rightarrow Cd(NO_3)_2(aq) + H_2S(aq)$
Acid–base reaction	$HA + B \rightarrow A + HB$ or $HA + NaOH \rightarrow NaA + H_2O$	$HCN(g) + NaOH(s) \rightarrow NaCN(s) + H_2O(g)$
Redox or oxidation-reduction reaction	Always involves a loss and gain of e^-	$SnO_2(s) + 2C(s) \rightarrow Sn(l) + 2CO(g)$
Combustion	$C_xH_y + O_2(g) \rightarrow CO_2(g) + H_2O(g)$	$2C_4H_{10}(l) + 13O_2(g) \rightarrow 8CO_2(g) + 10H_2O(g)$

changing as more information is obtained through research, so some misconceptions may be due to old ideas.

MISCONCEPTION: ELECTRONS IN ATOMS FOLLOW A CIRCULAR PATTERN. *Corrected Statement: Electrons do not follow circular patterns around the nucleus.* Explanation: Atoms are represented in comic books, the nuclear hazard symbol, and even some textbooks as electrons moving about the nucleus in a circular pattern like planets orbiting a sun. This is a simplified representation that was thought to explain electron behaviour in the early 1900s. We now know that electron movement is more complex and almost impossible to predict accurately.

MISCONCEPTION: THE NUCLEUS OF AN ATOM IS SIMILAR TO THE NUCLEUS OF A CELL. *Corrected Statement: The nucleus of an atom is completely unlike the nucleus of a cell.* Explanation: This is an unfortunate situation where the same word is used for two entirely different situations ("nucleus" comes from Latin *nucleus,* meaning "kernel or nut," and generally refers to a "central important unit"). The only similarity between the two types of nuclei is that they are inside something else, generally in or near the centre. The nucleus of an atom contains only protons and neutrons. The nucleus of a eukaryotic cell also contains protons and neutrons, but they are part of the atoms that form the molecules that are found in the membrane, proteins, and nucleic acids of which the cell nucleus is composed.

MISCONCEPTION: AN ELECTRON SHELL IS LIKE AN EGGSHELL THAT CAN BE TOUCHED OR BROKEN. *Corrected Statement: An electron shell has no physical reality. It cannot be touched or broken.* Explanation: *Electron shell* is a term used by some chemists to describe a region outside the nucleus where electrons can be found. Other chemists use the term *electron cloud.* Both of these terms are attempts by chemistry instructors to help students visualize a very abstract concept. In fact, while the energies that certain electrons possess can be calculated and predictions can be made about where those electrons might be located generally, there is no way to ascertain with any certainty where an electron is at any given time. Some terms that are also used to discuss electron movement and location are *atomic orbitals* and *electron energy levels.*

MISCONCEPTION: ELECTRONS BELONG TO A SPECIFIC ATOM. *Corrected Statement: Electrons do not belong to a specific atom.* Explanation: Electrons are governed by principles of energy and attractive and repulsive forces that cause them to associate with a certain nucleus for a period of time. Electrons are regularly lost to a particular atom, gained by another atom, or held in common between atoms. Electrons are usually found in an arrangement in either an atom or a molecule that allows it to be as stable (or low in energy) as possible.

MISCONCEPTION: IONIC BONDS ARE STRONGER THAN COVALENT BONDS, WHICH ARE BOTH STRONGER THAN INTERMOLECULAR FORCES. *Corrected Statement: Ionic compounds usually have higher melting points and heats of vaporization than covalent compounds.* Explanation: Because it is harder to melt ionic compounds than covalent compounds, people often assume that the bonds are stronger in ionic compounds than in covalent compounds. In reality, you cannot directly compare these forces because they are different types of forces. Ionic compounds are composed not of molecules but of ions that are attracted to one another. This is why the smallest unit of an ionic compound is called a formula unit, not a molecule. In an ionic compound, every ion is attracted to all the ions of opposite charge that surround it. For example, a sodium ion that is surrounded by chloride ions is equally attracted to all of the chloride ions around it. It is some of these bonds that are broken when an ionic compound is melted. On the other hand, in a covalent or molecular compound, the molecules are held together by intermolecular forces, which are quite weak, and when a covalent compound is melted it is only the intermolecular forces that have to be overcome. The molecules themselves stay intact. So when looking at the energy required to melt the different kinds of compounds, you are comparing ionic bonds to intermolecular forces, which are far weaker. This is why covalent compounds melt at much lower temperatures than ionic compounds.

It is almost impossible to make a general statement comparing ionic bond strength directly to covalent bond strength. For example, ionic bonds are generally easier to break than covalent bonds by dissolving a substance in water. So the best answer I can give you is that there is a wide range of both ionic bond strengths and covalent bond strengths, and statements that say one is stronger than the other are oversimplified.

 Go to **science3.nelson.com** for more examples of chemistry key concepts and downloadable tables and concept maps.

6.4 BIOLOGY BASICS

"Biology is a science of three dimensions. The first is the study of each species across all levels of biological organization, molecule to cell to organism to population to ecosystem. The second dimension is the diversity of all species in the biosphere. The third dimension is the history of each species in turn, comprising both its genetic evolution and the environmental change that drove the evolution. Biology, by growing in all three dimensions, is progressing toward unification and will continue to do so."

EDWARD O. WILSON

E. O. Wilson (2005). Systematics and the future of biology. *Systematics and the Origin of Species: On Ernst Mayr's 100th Anniversary, 102*(22–26), 1.

Brief History of Biology

Biology is the study of life—the study of organisms. If I were to pick a date in history when the study of life really started to take off, it would be when Robert Hooke and Antony van Leeuwenhoek developed and improved different types of microscopes in the 1600s. This is when bacteria, sperm cells, red blood cells, and microscopic protists were seen for the first time. This is also when it was discovered that living organisms are composed of cells. In the 1700s, Carl Linnaeus came up with a new way of categorizing organisms—a new taxonomy—and he painstakingly set out to label all species with scientific names. In the 1800s, great work was done on the cell theory, such as Louis Pasteur's work confirming that cells arise from pre-existing cells. It was also during the 1800s when Charles Darwin published *On the Origin of Species,* which detailed his theory of **evolution** by **natural selection**. Over the next 150 years after its publication in 1859, his theory would be accepted by scientists worldwide and become the foundational theory of biology. The 1900s were full of great discoveries in the realm of molecular biology, including the discovery of the structure of DNA in 1953. The Human Genome Project was completed in 2003, and now the genomes of more than 2000 organisms have been sequenced (~1500 of which are prokaryotic organisms). There is currently an "omics" revolution in biology, with an explosion of research into genomics, proteomics, transcriptomics, metabolomics, and more. Currently, projects are under way that use genomic technologies to identify which drugs will work best to treat individual breast tumours, to figure out why some trees succumb to pest infections and why others don't, to try to predict which people will succumb to certain diseases, and to learn why humans and chimpanzees look so different and succumb to different diseases despite having a genome that is 99 percent identical. (This list is just a very small sampling of current work in the field of genomics.)

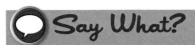

Say What?

BIOLOGY: The word *biology* comes from the Greek words *bios,* meaning "mode of life," and *-logia,* meaning "saying, speaking" or "a department of knowledge." Have you ever noticed that the root *bio-* is also in the word antibiotic, which means "against life"?

Go to **science3.nelson.com** for many more examples of current biology research.

 TOOL KIT

If you are interested in learning more about genomics projects, then I suggest you take a look at Genome Canada's website at **www.genomecanada.ca**.

Who's Who of Biology

FREDERICK BANTING (1891–1941). Banting, a Canadian physician, was one of the discoverers of insulin, for which he received the Nobel Prize in Physiology or Medicine in 1923. Although he shared the Nobel Prize with John Macleod, who was a supervisor at Banting's research institute, Banting shared his prize money with Charles Best, whom he credited with the co-discovery of insulin.

FRANCIS CRICK (1916–2004). An English molecular biologist who co-discovered the double-helical structure of DNA with James Watson, Crick wrote extensively on the "central dogma" of molecular biology—the idea that information travels one way in the cell from DNA to RNA to protein. The 1962 Nobel Prize in Physiology or Medicine was awarded to Francis Crick, James Watson, and Maurice Wilkins for their work on nucleic acids.

GEORGES CUVIER (1769–1832). A French naturalist and zoologist, Cuvier founded the discipline of vertebrate paleontology and proved that past organisms had gone extinct.

CHARLES DARWIN (1809–1882). An English naturalist who devised the theory of evolution by natural selection, Darwin outlined his theory in his 1859 book *On the Origin of Species*. The theory of evolution by natural selection is accepted as the core theory of biology.

ERASMUS DARWIN (1731–1802). Erasmus Darwin (the grandfather of Charles Darwin), a physician and naturalist, wrote about the idea of evolution. Although he did not describe natural selection, Darwin did discuss the idea of a single common ancestor for all species.

PAUL EHRLICH (1854–1915). A German scientist who invented the treatment of chemo-therapy and developed the first treatment for syphilis, Ehrlich received the Nobel Prize for Physiology or Medicine in 1908 for his work on the treatment of infectious diseases.

ALEXANDER FLEMING (1881–1955). A Scottish biologist who discovered penicillin, Fleming was awarded the Nobel Prize in Physiology or Medicine in 1945. The prize was shared with Howard Floery and Ernst Chain, the scientists who were able to figure out how to produce penicillin.

ROSALIND FRANKLIN (1920–1958). A British biophysicist who researched the molecular structure of DNA and RNA using X-ray diffraction, she produced data that were instrumental in the development of Watson and Crick's hypothesis of the structure of DNA. Had she been alive when Watson and Crick received their Nobel Prize in 1962, she would likely also have received it. However, Nobel Prizes were not usually awarded posthumously at that time.

STEPHEN JAY GOULD (1941–2002). An American evolutionary biologist and author, Gould came up with the theory of punctuated equilibrium, which proposes that evolution consists of relatively stable, long time periods that are "punctuated" by periods of rapid evolutionary change.

ROBERT HOOKE (1635–1703). An English scientist who made many contributions to both physics and biology, Hooke developed the law of elasticity (known as Hooke's law) and devised a type of compound microscope. Using this microscope, he became the first person to see plant cells, and he even came up with the term *cell* to describe them.

ROBERT KOCH (1843–1910). A German physician who isolated bacteria responsible for several different diseases, including anthrax, tuberculosis, and cholera, Koch came up with a method of confirming that a bacterium causes a specific disease. The method consists of analyzing four criteria (called Koch's postulates) that will identify a causal link between the bacterium and the disease.

CARL LINNAEUS (1707–1778). A Swedish naturalist who published a classification of all living things called *Systema Naturae*, Linnaeus is often called the "Father of Taxonomy" because he devised a system for naming and categorizing organisms. The naming method involved using two names—a Latin name for the genus, and then a common name (or shorthand name) for the species. This resulted in every organism being given a binomial (meaning "two names") label.

BARBARA MCCLINTOCK (1902–1992). An American geneticist who discovered the existence of transposons ("jumping genes") and discovered that genes have the ability to turn physical traits "on or off," McClintock earned the 1983 Nobel Prize in Physiology or Medicine for her work.

GREGOR MENDEL (1822–1884). An Austrian scientist who is considered to be the "Father of Genetics" due to his research into heritable traits in pea plants, Mendel's work forms the basis for the laws of Mendelian inheritance, which include the law of segregation and the law of independent assortment.

KARY MULLIS (1944–). An American biochemist, Mullis invented the polymerase chain reaction (PCR), for which he received the Nobel Prize in Chemistry in 1993. This prize was shared with Canadian researcher Michael Smith, who developed site-directed mutagenesis.

LOUIS PASTEUR (1822–1895). A French microbiologist who performed experiments that proved the "all cells from cells" hypothesis, disproving the hypothesis of spontaneous generation, Pasteur also created the first anthrax and rabies vaccines.

JONAS SALK (1914–1995). An American virologist, Salk developed the first vaccine for polio.

NIKOLAAS TINBERGEN ("NIKO") (1907–1988). A Dutch biologist and ethologist (someone who studies animal behaviour), Tinbergen, together with Konrad Lorenz, founded the field of ethology (also known as animal behaviour).

ANTONY VAN LEEUWENHOEK (1632–1723). A Dutch scientist, van Leeuwenhoek made hundreds of microscopes, and in turn made many microscopic discoveries. Some of these discoveries are bacteria, microscopic protists, and sperm cells.

ALFRED RUSSEL WALLACE (1923–1913). A British biologist, Wallace independently proposed the theory of evolution by natural selection at the same time as Charles Darwin.

JAMES WATSON (1928–). An American biologist who co-discovered the double-helical structure of DNA with Francis Crick, Watson was awarded the 1962 Nobel Prize in Physiology or Medicine, which was shared with Francis Crick and Maurice Wilkins for their work on nucleic acids.

ALFRED WEDENER (1880–1930). A German geophysicist, Wedener came up with the theory of continental drift.

EDWARD O. WILSON (1929–). An American biologist and environmental activist, Wilson founded the discipline of sociobiology.

Common Biology Abbreviations

In Box 6.10 I provide a list of common biology abbreviations. Some of these abbreviations overlap with the physics and chemistry abbreviations given earlier in the chapter.

BOX 6.10	Common Biology Abbreviations
A	adenine; adenosine; the letter abbreviation for the amino acid alanine
Ab	antibody
Ac	acetyl group
acetyl CoA	acetyl coenzyme A
ADA	adenosine deaminase
ADH	alcohol dehydrogenase
ADP	adenosine 5'-diphosphate
Ag	antigen
AIDS	acquired immune deficiency syndrome
Ala	alanine residue (also identified as A)

Continued

AMP	adenosine 5'-monophosphate
ANOVA	analysis of variance
Arg	arginine residue (also identified as R)
Asn	asparagine residue (also identified as N)
Asp	aspartic acid residue (also identified as D)
ATP	adenosine triphosphate
ATPase	adenosine triphosphatase
β-gal	β-galactosidase
BAC	bacterial artificial chromosome
BLAST	Basic Local Alignment Research Tool
bp	base pair
C	cytosine or cytidine; also the one-letter code for the amino acid cysteine
cal	calorie (4.18 J)
cDNA	complementary deoxyribonucleic acid
CFU	colony-forming unit
CGH	comparative genome hybridization
Ci	curie
cM	centimorgan
cpm	counts per minute
cys	cysteine residue (also identified as C)
Da	dalton
DNA	deoxyribonucleic acid
DNAse	deoxyribonuclease
dNTP	deoxynucleoside triphosphate
ds	double-stranded
EGF	epidermal growth factor
ER	endoplasmic reticulum
FACS	fluorescence-activated cell sorting
FISH	fluorescence in situ hybridization
fMet	formyl-methionine
FRET	fluorescent resonant energy transfer

Continued

G	guanine or guanosine; one-letter code for the amino acid glycine
G	Gibbs free energy of a system
GFP	green fluorescent protein
Glc	glucose residue
Gln	glutamine residue (also identified as Q)
Glu	glutamic acid residue (also identified as E)
Gly	glycine residue (also identified as G)
GTF	general transcription factor
GTP	guanosine 5′-triphosphate
H	enthalpy of a system
hGH	human growth hormone
His	histidine residue (also identified as H)
HIV	human immunodeficiency virus
HLA	histocompatibility locus antigen
hsiRNA	heterochromatic short interfering RNA
HSV	herpes simplex virus
I	inosine
Ig	immunoglobulin
Ile	isoleucine residue (also identified as I)
J	joule
K	degrees kelvin (absolute temperature)
kb	kilobase
kcal	kilocalorie
K_d	dissociation constant
kDa	kilodalton
LDL	low-density lipoprotein
Leu	leucine residue (also identified as L)
Lys	lysine residue (also identified as K)
LTR	long terminal repeat
m	metre
M	relative molecular mass
mAb	monoclonal antibody

Continued

Mb	megabase; megabyte
MCS	multiple cloning site
Met	methionine residue (also identified as M)
MHC	major histocompatibility locus
mol	mole
mp	melting point
mRNA	messenger ribonucleic acid
NAD	nicotinamide adenine dinucleotide
NAD^+	nicotinamide adenine dinucleotide (oxidized)
NADH	nicotinamide adenine dinucleotide (reduced)
$NADP^+$	nicotinamide adenine dinucleotide phosphate (oxidized)
NADPH	nicotinamide adenine dinucleotide phosphate (reduced)
NCBI	National Center for Biotechnology Information (U.S.)
NCI	National Cancer Institute (U.S.)
neo	neomycin gene (used as a selectable marker)
NK	natural killer cells
NMR	nuclear magnetic resonance
nt	nucleotide
oligo	oligonucleotide
OMIM	Online Mendelian Inheritance in Man
ORF	open reading frame
ori	origin of replication
PCD	programmed cell death
PCR	polymerase chain reaction
Phe	phenylalanine residue (also identified as F)
P_i	inorganic phosphate
ppm	parts per million
Pro	proline residue (also identified as P)
Pu	purine
Py	pyrimidine
RBC	red blood cell

Continued

RBS	ribosome-binding site
RE	restriction endonuclease
RFLP	restriction-fragment-length polymorphism
RNA	ribonucleic acid
RNAse	ribonuclease
rRNA	ribosomal ribonucleic acid
RT	reverse transcriptase
RT-PCR	reverse transcriptase polymerase chain reaction
S	entropy of a system
S	svedberg sedimentation unit
siRNA	short interfering ribonucleic acid
snRNA	small nuclear ribonucleic acid
SRP	signal recognition particle
ss	single-stranded
STS	sequence tagged site
T	thymine or thymidine; one-letter code for the amino acid threonine
Taq	thermus aquaticus DNA (which is used as a polymerase)
TBT	TATA-binding protein
TCR	T cell receptor
TF	transcription factor
TGN	trans-Golgi network
Thr	threonine residue (also identified as T)
tRNA	transfer ribonucleic acid
Trp	tryptophan residue (also identified as W)
Tyr	tyrosine residue (also identified as Y)
U	uracil
UTR	untranslated region
UV	ultraviolet
Val	valine residue (also identified as V)
WBC	white blood cell
WT	wild-type

Different Branches of Biology

I describe different branches of biology in Box 6.11.

BOX 6.11   Different Branches of Biology

Anatomy and physiology: The study of how living organisms and their components (i.e., organs) function, with an emphasis on the relationship between structure and function.

Botany: The study of plants.

Cell biology: The study of cells and how they function.

Conservation biology: The study of the protection and conservation of organisms, ecosystems, and environments.

Developmental biology: The study of developmental processes involved in the growth of the zygote to a full organism.

Ecology: The study of how organisms interact with each other and how they interact with the environment.

Embryology: The study of how an embryo develops.

Entomology: The study of insects.

Evolutionary biology: The study of how species have evolved over time.

Genetics: The study of genes and how they are inherited.

Microbiology: The study of microorganisms (such as viruses, protists, and bacteria).

Molecular biology: The study of biology at the level of DNA, RNA, and proteins.

Mycology: The study of fungi.

Paleontology: The study of fossils.

Systematics: The study of classifying organisms.

Zoology: The study of animals.

Biology can be organized into different branches of study, but it can also be organized into a hierarchy. This hierarchical method of organization goes from the smallest and least complex component of biology—the atom, to the largest and most complex—the biosphere. Box 6.12 leads you through the levels of biological hierarchy and provides corresponding examples.

BOX 6.12   The Hierarchy of Biology

Hierarchy Level	Example	Properties Arising at This Level of Structure
Atom	Hydrogen	Electronegativity; ionization; radioactivity; and features of electron configuration (energy levels, valence) occur at this level.
Molecule	Water (H_2O)	The ability to form particular types of bonds (e.g., hydrogen, ionic, covalent, van Der Waals interactions) and the beginnings of 3-D shape occur at this level.
Macromolecule	DNA	With large molecules arising from consistent monomers, simple and repeatable molecules with specific functions are better supported.
Organelle	Nucleus	Groups of macromolecules can contribute to partitions (e.g., membranes) or form clusters of specific metabolic function (e.g., enzymes within a lysosome).
Cell	Neuron	With a complete complement of organelles, a structure that exhibits ALL the characteristics of life can arise. There is no known entity below the cellular level that demonstrates all the properties of life.

Continued

Tissue	Neural tissue	With a group of similar cells working together, new features can be achieved. A single neuron doesn't think, but a collection of neurons can form a decision-making body such as a neuron.
Organ	Brain	With the support of several cell types, an entity can be created that is capable of more complex activity, such as the formation of memories.
Organ system	Nervous system	Connection of and communication between organs all involved in a similar process can coordinate activities effectively (such as proper progression of food through the mouth, pharynx, esophagus, stomach, and intestines).
Organism	Human	All the organ systems together are required to create an autonomous unit that isn't physically and permanently connected to other units.
Population	A group of humans living in a town	Groups of individuals may form the capacity for sexual reproduction, which is a powerful evolutionary tool.
Community	The humans and other animals that live in an alpine town	Having plants and animals in a localized geographical region allows interflow of energy and organic molecules for the first example of a self-perpetuating system.
Ecosystem	All the organisms and environmental factors in an alpine town	Local environments vary in temperature, rainfall, terrain, and other abiotic factors that shape the organisms within it through natural selection. Organisms are a part of this because they, too, exert shaping forces on their neighbours.
Biosphere	Earth	Different ecosystems are not isolated, so this level of complexity encompasses the incredibly complex influences different ecosystems have on each other as well as large-scale changes that occur to all of them at once (such as climate change).

The reason a hierarchy of biology is so important is that it helps you understand how "emergent properties" work. The term *emergent properties* describes new attributes or abilities that arise as a structure becomes more complex. At the atomic level of structure, which is very simple, the atoms can form bonds and rearrange electrons to store and release energy. However, when you put two atoms together to form a molecule, the types of bonds that are formed can make the molecule more soluble (if the bonds are ionic or polar covalent) or less soluble (if non-polar covalent). This means that at the molecular level, molecules can spontaneously form sheets, like a membrane. However, a membrane does not exhibit all the properties of life. It is only at the cellular level of structure that there is sufficient complexity to reasonably suggest that the object has the properties of life. Look over the list in Box 6.12 and ask yourself what new properties each level has that makes it distinct from the items lower down in the hierarchy.

 Go to **science3.nelson.com** for more examples of biology hierarchies.

Key Biology Themes/Concepts

"Nothing in biology makes sense except in the light of evolution."

THEODOSIUS DOBZHANSKY, 1973

KEY BIOLOGY THEME: EVOLUTION BY NATURAL SELECTION IS THE CENTRAL THEME OF BIOLOGY. The process of **evolution** by natural selection is responsible for both the unity that connects all living things on Earth and the diversity that is seen across living organisms. Evolution is a change in the alleles of a given population over time. The major mechanism causing evolution is *natural selection,* although genetic drift, mutation, and artificial selection can also cause evolution (see Box 6.13).

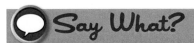

EVOLUTION: The word *evolution* means a change in the genetic makeup of a population over time, and comes from the Latin word *evolvere,* meaning to "unroll."

What Can Cause Evolution?

1. **Natural selection:** The major mechanism of evolution; a process by which individuals with heritable traits that are beneficial in a specific environment will produce more offspring, and the alleles responsible for the trait will be found in more members of that population.

2. **Genetic drift:** A random change in allele frequencies.

3. **Mutation:** A random change in DNA sequence.

4. **Artificial selection:** Deliberate selection (by humans, and even other animals) of traits and corresponding genes in a specific plant or animal.

Charles Darwin proposed the theory of evolution by natural selection in 1859. He proposed that all organisms contain heritable variations, and that these heritable variations will lead to better survival and reproduction depending on the environmental conditions. If an individual organism survives and demonstrates good health and other characteristics that allow it to mate more successfully, then that organism will probably pass on more of its genes (and the traits encoded by those genes) to the next generation. Increased genetic diversity will increase the likelihood of survival for individuals in a given population if the environment changes, because traits that were useful in a previous environment may be less desirable in the new one. Diversity means that there are other options available for selection. Organisms that produce more viable offspring than other organisms are said to have higher *fitness*.

 TOOL KIT

The Understanding Evolution website (**www.evolution.berkeley.edu**) developed by the University of California at Berkeley is an excellent resource that contains many study aids and tools to help you learn about evolution.

All organisms on this planet, whether living or extinct, are connected to a common ancestor that existed approximately 4.6 billion years ago. We know that they are connected because all organisms on Earth have several things in common. These things include (1) a universal genetic code and (2) shared metabolic pathways.

Organisms on Earth can be divided into three domains: Bacteria, Archaea, and Eukarya. The domains Bacteria and Archaea consist of prokaryotes (unicellular organisms that usually lack membrane-bound organelles and have no defined nucleus), and the domain Eukarya consists of eukaryotes (organisms that can be unicellular or multicellular and have membrane-bound organelles, and in particular a membrane that surrounds their chromosomes to form a distinct nucleus). Organisms in each domain share common characteristics, providing further evidence of common ancestry. For example, eukaryotes all have a nucleus that contains linear chromosomes, membrane-bound organelles and an endomembrane system, a cytoskeleton, and a double phospholipid bilayer as a plasma membrane.

KEY BIOLOGY THEME: ALL LIFE IS MADE UP OF CELLS. The cell theory consists of two main tenets: (1) All life is made up of cells, and (2) cells arise from pre-existing cells. Cells can be either prokaryotic or eukaryotic. Prokaryotic cells lack organelles but still have a highly organized structure within their membranes. Eukaryotic cells have organelles, and each organelle has specific structure and function.

Go to **science3.nelson.com** for an exercise that will help you learn/ review the structures and functions of the various organelles.

Prokaryotic cells divide by the process of *binary fission,* wherein one cell splits into two or more cells. This is a form of *asexual reproduction.* Eukaryotic cells can divide via mitosis or meiosis. **Mitosis** is a process of cell division that produces daughter cells that are a genetic match to the parent cell, whereas **meiosis** produces daughter cells that are not a genetic match to the parent cell.

There are three instances when mitosis will occur: (1) to allow the organism to grow, (2) to heal a wound or injury, and (3) to reproduce (asexual reproduction). Meiosis also occurs to reproduce (sexual reproduction) and, in humans and most animals, is the process that is used to make eggs and sperm. The daughter cells have half the chromosomes of the parent cells, and the chromosomes are genetically different due to recombination events. Mitosis produces two daughter cells from one parent cell, and the daughter cells are identical to each other and to the parent. However, meiosis produces four daughter cells from one parent cell, and the daughter cells contain half as much genetic material as the parent cell and are genetically different from each other and from the parent.

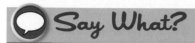

Say What?

MITOSIS: The word *mitosis* comes from the Greek word *mitos,* meaning "thread," referring to the threadlike appearance of condensed chromatin at the beginning of mitosis.

MEIOSIS: The word *meiosis* comes from the Greek word *meioun,* which means "to lessen." Recall that meiosis is a reduction division, wherein the resulting daughter cells have half the genetic content of the parent cells.

KEY BIOLOGY THEME: ORGANISMS NEED ENERGY TO SURVIVE. Organisms can be categorized based on how they get energy: Autotrophs capture free energy through photosynthesis or chemosynthesis, whereas heterotrophs eat other organisms to get energy. (These other organisms could be either other heterotrophs or autotrophs. The key point is that heterotrophs capture energy by consuming carbon compounds, such as carbohydrates, proteins, and lipids.) The process of photosynthesis, performed by autotrophs, uses chlorophyll to trap free energy from sunlight and makes carbohydrates from CO_2. (The process of chemosynthesis gets energy from inorganic chemicals.) The process of cellular respiration then breaks down the carbohydrates to get the energy back and produces free energy carriers such as ATP. These free energy carriers are then used to power cellular work, including metabolic reactions. The processes of photosynthesis and cellular respiration are linked and cyclical—they are interdependent. Note that if there is no oxygen available for cellular respiration, then fermentation can occur, but it is a much less efficient method of getting energy from carbon-based macromolecules. The formulas for photosynthesis and cellular respiration are as follows:

Photosynthesis: $6CO_2 + 6H_2O + \text{Light Energy} \rightarrow C_6H_{12}O_6 + 6O_2$

Cellular Respiration: $C_6H_{12}O_6 + 6O_2 \rightarrow 6CO_2 + 6H_2O + \text{Energy}$

KEY BIOLOGY THEME: ORGANISMS NEED TO MAINTAIN HOMEOSTASIS IN ORDER TO SURVIVE. Organisms need to closely regulate their internal environment, such as temperature and metabolism. Feedback mechanisms enable them to respond to any environmental changes: Positive feedback loops (such as the beginning of labour during childbirth) are used to intensify and strengthen responses, while negative feedback loops (such as maintenance of temperature) are used to adjust a condition and then stop the response once the goal has been achieved. Organisms are able to maintain **homeostasis** via behavioural mechanisms (such as moving to the shade to cool down) and physiological mechanisms (such as sweating).

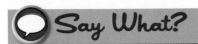

HOMEOSTASIS: The word *homeostasis* comes from the Greek words *homoios*, meaning "same," and *stasis*, meaning "standing still," or "stationariness." So homeostasis means "the maintenance of a stable state."

KEY BIOLOGY THEME: LIFE IS ONLY POSSIBLE DUE TO THE UNIQUE PROPERTIES OF WATER. Life is completely dependent on water. Not only are organisms made up predominantly of water, but if it weren't for the molecular shape of water and its high polarity, which gives it the ability to form hydrogen bonds, life wouldn't exist as it does today. Its ability to readily form hydrogen bonds makes water a very good solvent (meaning things get dissolved in it, particularly molecules with charges associated with them).

KEY BIOLOGY THEME: LIVING ORGANISMS HAVE A UNIVERSAL GENETIC CODE, AND INFORMATION FLOWS FROM THIS GENETIC CODE THROUGH TO RNA AND THEN TO PROTEINS. DNA contains all the information a cell needs to survive, grow, and reproduce. DNA consists of just four different nucleotides—A, G, C, and T—and each nucleotide consists of a sugar, a phosphate, and a nitrogenous base. Nucleotides are linked together through phosphodiester bonds and stabilized by hydrogen bonds to form a double-helix DNA molecule. These four nucleotides are common to all types of life on the planet. These four bases code for 20 different amino acids. In order to make proteins, the cell transcribes DNA into RNA, and then (often, but not always) translates the RNA into protein. If the DNA changes (acquires a mutation), then the protein structure may change. This means that a change at the level of DNA (the genotype) can be seen at the level of the protein (which is mostly responsible for phenotype). This is why different organisms have different appearances—because they have differences in their DNA that in turn cause proteins to be made differently to result in a different phenotype.

Common Biology Misconceptions

MISCONCEPTION: INDIVIDUAL ORGANISMS CAN EVOLVE. *Corrected Statement: Populations can evolve, but individual organisms cannot (Alters & Nelson, 2002).* Explanation: Recall that evolution means "change over time." The time that is referred to in that phrase is the generation time. An individual's genes do not change during its lifetime. (Exceptions are the accumulation of mutations that could, for example, cause cancer, and the genetics dictating the construction of antibody molecules. However, these mutations usually occur in somatic cells that are not passed on to offspring.)

MISCONCEPTION: THE DNA DOUBLE HELIX IS A MOLECULE. *Corrected Statement: The DNA double helix is two macromolecules that are held together with hydrogen bonds.* Explanation: This misconception is perpetuated in most introductory biology books. Some authors try to hedge their bets by calling the double helix a "strand of DNA," but many books insist that it's "a molecule." A molecule is a group of atoms held together by covalent bonds, and the two DNA polynucleotides are definitely NOT covalently linked! Otherwise they could not separate during transcription or replication! Each strand of DNA is in fact a macromolecule. (A macromolecule is a very large molecule made up of subunits, or monomers, that are bonded together.) The two macromolecules of DNA associate through hydrogen bonds between AT and GC nucleotides, which give them particular stability. However, if hydrogen bonds were sufficient to qualify the bonded entities as a molecule, the fluid in a glass of water could be considered to be a single molecule.

MISCONCEPTION: MUTATIONS OCCUR IN RESPONSE TO AN ENVIRONMENTAL CHANGE. *Corrected Statement: Mutations occur at random and exist in a population. If the environment changes, then a mutation that was previously neutral may become beneficial and increase in frequency in a population due to differential survival of individuals with the mutation.* Explanation: An example of this misconception is, "the environment got hotter, so mutations occurred to cause the animal's skin to turn white to reflect the sun's energy" (Alters & Nelson, 2002). Mutations do not occur just because the environment has changed and they "need" to occur in order for an organism to survive. The occurrence of mutations is a random process, and when mutations do occur they may be beneficial, neutral, or deleterious.

MISCONCEPTION: HUMANS EVOLVED FROM CHIMPANZEES. *Corrected Statement: Humans and chimpanzees have a common ancestor.* Explanation: Humans did not evolve from chimpanzees. Rather, humans and chimpanzees have a common ancestor that lived approximately 6 million years ago. This common ancestor diverged into many species, two of which eventually led to the evolution of humans and chimpanzees.

MISCONCEPTION: MEIOSIS ALWAYS RESULTS IN GAMETES. *Corrected Statement: Meiosis results in haploid cells.* Explanation: In humans and most animals, haploid cells always form gametes. However, in plants and many protists (single-celled eukaryotes), meiosis creates haploid cells that divide mitotically. Only a few of these haploid cells will be gametes, and only if the organism is pursuing a sexual life cycle. Some organisms don't, by the way!

MISCONCEPTION: ONLY DIPLOID CELLS DIVIDE BY MITOSIS. *Corrected Statement: All eukaryotic cells are potentially capable of mitotic division.* Explanation: Many students try to group things based on simple characteristics. Being diploid or haploid doesn't affect the necessity for cell division. Since most biology books communicate human cell features, one might think that haploid cells don't divide through mitosis—and they don't, in humans! But many organisms are multicellular haploids—bee drones (males), for example, are haploid. They form by mitotic divisions of an unfertilized ovum from the queen bee. The males also form sperm, of course—but since they're already haploid, the sperm are not formed through meiosis in the male!

 Go to **science3.nelson.com** for more examples of biology key concepts and downloadable tables and concept maps.

6.5 CONCLUSION

You likely already studied the majority of the content in this chapter in high school. Your courses at college or university will go into much greater depth than what this chapter covers. So, how will you be able to study effectively and manage your time well so that you can comprehend and remember all of the wonderful things you are going to learn about science? For that, let's turn to Chapter 7.

FURTHER READING

Alters, B. J., & Nelson, C. E. (2002). Perspective: Evolution in higher education. *Evolution, 56*(10), 1891–1901.

Lazarowitz, R., and Lieb, C. (2006). Formative assessment pre-test to identify college students' prior knowledge, misconceptions and learning difficulties in biology. *International Journal of Science and Mathematical Education, 4,* 741–762.

Malin, J. M. "International Year of Chemistry—2011 Chemistry—our life, our future."

Miller, J. C. (2002, April). College students' beliefs, alternate conceptions and understanding of the theory of common descent. Paper presented at the Annual Meeting of the National Association for Research in Science Teaching (NARST). New Orleans, LA.

Stern, L., and Mokady, O. (2004, April). Will dinosaurs ever appear again? University biology students' conceptions of determinism in nature. Paper presented at the Annual Meeting of the National Association for Research in Science Teaching (NARST). Vancouver, BC.

Treagust, D. F. (1988). Development and use of diagnostic tests to evaluate students' misconceptions in science. *International Journal of Science Education, 10*(2), 159–169.

Study Strategies

"Proper preparation and planning prevents piss poor performance" (also known as the 7 Ps).

BRITISH ARMY SAYING

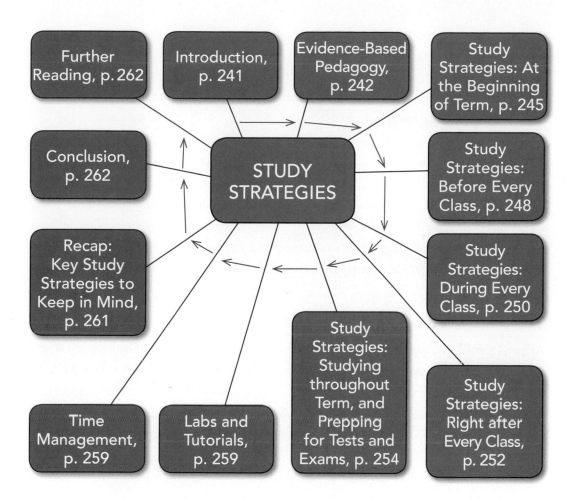

7.1 INTRODUCTION

No matter what topic you are studying, be it gene regulation, covalent bonding, or atomic structure, there are several study strategies you can use to learn the topic well. And by "learn the topic well" I mean that you will achieve a deep level of understanding and be able to **recall** this information for a long time after you finish the course. More important, you'll be able to apply concepts to new situations, a skill that is much more valued by educators and employers than rote memorization. In this chapter, I'll guide you through some study strategies that you'll be able to apply to all of your university courses.

First, think about how you study right now. You might have already been strategic and chosen one place to study and have it set up with a glass of water, pencils, and tissues nearby. Perhaps you focus on one subject for a long period of time until you have mastered it, and maybe you do sample tests just before exams. Would you be surprised if I told you that these three strategies have been shown NOT to be the most effective? The evidence, which comes from well-designed studies, shows that varying your study location, switching between study topics rather than focusing on one subject for a long time, and testing repeatedly throughout term are more effective. I'll get to these strategies in a few pages, but first I want to talk about evidence-based pedagogy and the research that is currently being done on how students learn.

7.2 EVIDENCE-BASED PEDAGOGY

As you learned in Chapter 1, evidence-based pedagogy is the science of how people learn. Many researchers are actively involved in analyzing the most effective ways to teach students and the most effective ways for students to study. (For a sample, see Bjork, 1994; Haak et al., 2011; Roediger et al., 2011; and Tomarek & Montplaisir, 2004.) Two of these researchers are Dr. Henry L. Roediger, III, who is a Professor of Psychology and runs the Memory Lab at Washington University in St. Louis, and Dr. Robert Bjork, a Cognitive Psychology Professor Emeritus at the University of California, Los Angeles, and a leading researcher on memory and learning. I was at a seminar given by Dr. Bjork a few years ago, during which he said that students need to "learn how to learn." I thought at the time that universities should be placing a greater emphasis on teaching evidence-based learning strategies to students in their first year. As a student, if you learn to identify the study strategies that work (and those that don't), then you'll be far more likely to excel at your current courses and leverage those great skills as you work your way through later, more challenging courses. That's what this chapter is all about—the best ways to go about learning the material from your courses.

The advice in this chapter is based on scholarly studies about how students learn—by reading this chapter, you will learn how to formulate an effective study strategy. If you want to know the nuts and bolts behind the **cognitive psychology** and **Scholarship of Teaching and Learning (SoTL)** studies that have led to this advice, then please visit the website associated with this book.

 Go to **science3.nelson.com** for expanded descriptions of the cognitive psychology and SoTL studies behind the study strategy advice given in this chapter.

So what has some of the SoTL research shown? Two areas of SoTL research that have received a lot of attention are **active learning** and **problem-based learning**, and many university-level courses implement these strategies in their course structure.

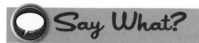 **Say What?**

ACTIVE LEARNING: A catch-all term for instructional strategies that get students actively involved in learning course material. Examples of active learning techniques include personal response systems (clickers), the think-pair-share technique, and the "learning by teaching" technique.

PROBLEM-BASED LEARNING (PBL): An instructional strategy in which students learn about a topic through the process of solving a difficult and multidimensional problem on that topic. A key aspect of PBL is that the problem should be realistic, meaning that people would actually need to solve that problem in everyday life.

Active learning involves the use of instructional strategies that move away from the "sit in your seat and be quiet while I lecture to you" method, toward the incorporation of techniques that get students actively involved in the learning process (reviewed in Gardner & Belland, 2012, and Michael, 2006). Let's take a look at three such techniques. (1) The **"think-pair-share" technique** involves posing a question or problem to students, having them *think* about it for a minute, then *pair* with a partner to discuss the problem further, followed by *sharing* their solution with the class. (2) **Personal response systems**, also known as **clickers**, are handheld devices (similar to a remote control) that let you enter an answer to a question posed by your professor, which then gets received by the professor's computer. Sometimes professors will show the class the outcome of a given question (e.g., a professor might ask the class a multiple choice question, then allow the class to answer the question using their clickers, and then show the range of responses back to the class). The incorporation of clickers into lectures has been shown to increase student engagement. (3) The **"learning by teaching" technique** involves any scenario whereby students are required to teach course content to each other. This technique can be combined with many other active learning techniques. For example, in some of my classes, I get students to create demonstrations of key concepts, then show these demonstrations to other students, and then answer questions from their fellow students. Active learning techniques can be applied to any topic in the science curriculum and have been shown to be effective in increasing student ability at science process skills such as hypothesis testing and statistical analysis (Goldstein & Flynn, 2011).

 Go to **science3.nelson.com** for videos showing how different active learning techniques are incorporated into first-year science classes.

Problem-based learning (also known as PBL) is a teaching method whereby teams of students learn course content by solving challenging problems. Curricula that incorporates PBL into first-year health sciences courses have been shown to lead to higher student grades and increased student enjoyment of the course (Assalis et al., 2012). PBL curriculum has been incorporated into many different science courses and is also used in medical education. McMaster University has been using a PBL curriculum in its medical school since 1969 (Lee & Chiu, 1997).

 Go to **science3.nelson.com** for sample PBL question sets for biology, chemistry, and physics classes.

If you want to find out what your professors are learning about evidence-based learning, take a look at some science education journals. Here is a small selection of journals I routinely access when reading about pedagogical research:

Bioscience Education (www.bioscience.heacademy.ac.uk/journal/)

CBE Life Sciences Education (www.lifescied.org/)

Journal of Biological Education (www.tandfonline.com/toc/rjbe20/current)

Journal of Chemical Education (pubs.acs.org/journal/jceda8/)

Journal of Research in Science Teaching (onlinelibrary.wiley.com/journal/10.1002/%28ISSN%291098-2736)

Physics Education (iopscience.iop.org/0031-9120/)

7.3 STUDY STRATEGIES: AT THE BEGINNING OF TERM

Science is a topic that builds on prior knowledge, so it is very difficult to catch up to course content if you fall behind. Because of this, you should make sure to get off on the right foot and start your study strategy even before classes start.

Where Will You Study?

One of the first things you have to decide is where you will study. You may have been told by previous teachers or by other study guides that the best thing to do is to choose one study location and always study there. However, this actually isn't the most effective approach. Research shows that varying your study location is actually more effective than studying in one place and will result in higher test scores (Cepeda et al., 2006). (The exception to this rule is if your exam is going to be in the same place as you study. If that's the case, then studying in that place tends to result in higher scores.)

STRATEGIES THAT WORK

Vary your study location rather than studying in just one place.

The Course Syllabus

The next action item of your study strategy is to get a copy of the course **syllabus** (sometimes also referred to as a course outline). You can usually get a copy from the registrar's office or through your university's website or **course management system** (such as Blackboard).

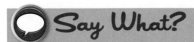

SYLLABUS: Syllabus comes from the Latin word *syllabus*, meaning "list." Although the term implies a list of course topics, the syllabus document usually contains much more information, such as late policies, office hours, and sometimes ancillary materials that can help you succeed in the course.

The course syllabus is usually posted online a few days before the start of classes, although some professors will hand it out only on the first day of class. This is a document that you should read from cover to cover and take notes on. Based on the information in the syllabus, be sure to do the following:

• Confirm the time and location of lectures, labs, and/or tutorials.

• Buy the textbook.

• Check to see if the recommended (but not required) books are available in the library. Some professors will refer to recommended books regularly, and some rarely. If you are wondering whether or not to buy books from the recommended list, I suggest you contact your professor directly via email to ask for his or her input.

• Review the course outline and the corresponding reading list.

- Review the learning outcomes of the course (more on this below).

- Acquaint yourself with all course policies. Two of the questions you should be able to answer are: What is the penalty for late assignments? (Not that you will get a penalty, because you will be so well organized and submit everything on time this term!) Does the course use plagiarism detection software such as TurnItIn?

- Find out what online course management system the course uses. Many courses use one, such as Blackboard or Desire2Learn. In some courses, you will be expected to access the system daily, whereas other courses may have no expectations. Be sure to find out how your professor will use the system (such as posting lecture notes, leading a discussion forum, or posting student grades), and be sure to ask the professor how often she or he expects you to access the system.

- Acquaint yourself with the course supports the professor has outlined in the syllabus. Course supports may include (but are not limited to) professor office hours, teaching assistant (TA) office hours, librarian support, online discussion forums, and posted sample midterm and final examinations.

 TOOL KIT

You should consider the course management system (such as Blackboard or Desire2Learn) one of the tools in your tool kit that will help you succeed. Be sure to get well acquainted with the system. Most libraries offer tutorials on how to use the online system, and several of these tutorials may be in video form on the university's website (depending on your institution). Note that every time you access the course management software, your activity is logged, and your professor can, if he or she chooses to, tell what tools you use (such as the discussion board, or the lecture slide downloads).

Rather than being something that you read through once and then abandon, the course syllabus is a document that you will want to access frequently throughout term. So always keep it close by, perhaps at the front of your course binder.

DANGER ZONE

Before you email your professor with a question, double-check the syllabus to make sure the answer isn't on it. Every term, professors receive emails from students with questions that could be answered from the syllabus. This is not the type of impression (i.e., that you are someone who doesn't read the syllabus) that you want to make on your professor at the start of term.

Learning Outcomes

Many professors include a list of learning outcomes for the course on the syllabus. This is a list of things you should be able to do on finishing the course. Make sure you refer back to these every week, and ask yourself if you have achieved them. Learning outcomes can also make excellent practice exam questions. For example, a learning outcome for a first-year biology course may be, "The student will be able to solve population genetics problem sets using the Hardy–Weinberg equation." If that is one of the learning outcomes on your syllabus, then make sure you practise these types of problems before writing any tests or exams.

7.4 STUDY STRATEGIES: BEFORE EVERY CLASS

I find that most of my students focus so much on what to study after class that they often forget to allocate some time to preparing for class. I encourage you to spend about 1 hour preparing for each class, so that you arrive to class with some advance understanding of the lecture material. This will allow you to *listen* during class (more on that in a few pages), rather than just writing down everything you hear. And since we're talking about the amount of time spent out of class studying, I should emphasize here that I advocate for 2 to 2.5 hours of out-of-class time for every hour in class. (This includes both preparation and review time, but does not include time spent on assigned homework, such as lab reports.) This varies by class, so be sure to check with your professors and ask them what their expectations are.

Before every class, be sure to:

- Check the syllabus to see what the class will be on.

- Read the relevant material in the textbook (see Box 7.1).

BOX 7.1   How to Use Your Textbook

Your textbook is not a list of information that you need to memorize. Rather, it is an excellent resource that can guide you in learning the course information. Think of it as just that—a guide. It is impossible for you to memorize everything in the textbook (although many students will try).

The following list contains suggestions on how you can get the most out of your textbook:

- Focus first on the key concepts, rather than the small details. Most textbooks contain a summary of key concepts at the beginning and end of each chapter. As you read through the text, be sure to ask yourself how the small details relate to the larger key concepts in the chapter. When you finish a chapter, you should be able to answer this question: "What are the primary concepts of this chapter?"

- When you have finished reading a page of the textbook, ask yourself if you understand the material you just read. One way to find out if you understand it is to think of a test question that the professor would ask on that material, and then provide an answer. For example, if you have just read a page on DNA replication, a question could be, "What would happen if helicase were not working properly?" The answer would be "DNA replication wouldn't proceed because the double helix would not get unwound by helicase."

- Focus on the figures. If you are able to describe a figure after covering the figure legend, then it shows you have some understanding of the concept that the figure demonstrates.

- Do the questions at the end of the chapter. I suggest doing them in a three-step process: (1) Do them in a testlike format (without looking back through the chapter), then (2) go back through the chapter and expand on your answers, and then (3) check the answer guide that comes with the textbook.

Note: Many textbooks come with an online supplement resource that may contain sample tests, flashcards, animations, and other exercises. If this resource comes with your textbook, then you should make use of it throughout term.

- If they are available, download the lecture notes and read them.

- Remember to study the word roots you learned in Chapter 4. Using word roots and context clues will help you prepare appropriately for class.

- Do a pretest for the upcoming material. Doing pretests has been shown to improve later learning (Kornell, Hays, & Bjork, 2009). If you don't have a pretest, you could read through the chapter and, if one is available from your textbook's website, do an online test as if it were a pretest to check your learning.

7.5 STUDY STRATEGIES: DURING EVERY CLASS

If you have followed the suggestions above and have come prepared to class, then you'll be able to actively listen and participate in class. You'll also be able to ask insightful questions, since it won't be the first time you're seeing the material.

For every class, be sure to:

- Go to class. There is a strong correlation between class attendance and student performance/ academic success (Moore et al., 2003; Newman-Ford et al., 2008; Obeidat et al., 2012).

- Not multitask during the lecture. This means do not check your phone, do not check Facebook, and don't play video games. Some students think that they are good at multitasking—but research has shown that you cannot multitask and learn class material effectively.

- Take notes, but do NOT write down everything the professor says. You aren't just a scribe; you're there to consider what's being said and form new ideas and understanding. Excessive writing cuts into that process. Some cognitive psychology research has indicated that waiting until after the lecture to write down your notes will result in better learning. Consider doing both—write down important notes during the lecture, and then enhance these notes after the lecture. For more advice on taking good notes, see Box 7.2.

- Use class time wisely. This is 1 hour out of your day when you won't have other distractions. Don't put off learning the material until later—focus on learning it now. I find that many students take notes in class with the intent of studying the notes later. Change your focus to learning in class.

BOX 7.2 How to Take Good Notes

Do NOT:

- Write down absolutely everything the professor says.

- Record the lecture, and then write down everything the professor said. Note that it is actually illegal to record the lecture without the instructor's permission. (Even if you get the instructor's permission, you can only use the recording for personal use; you are not allowed to distribute it, not even to other students in the class.)

Do:

- Come to class prepared with either printed lecture notes if they are provided by the instructor or some notes you made earlier based on your pre-reading.

- During the lecture, add to these notes when the professor says something that you don't recall from the pre-reading.

- Keep a running list of terms you don't understand in the margin of your notes, and be sure to look them up after class.

- Try different forms of note taking. For example, some students may find it helpful to take their notes in table format, like this:

Course: BIO100

Date: Sept. 18

Lecture Topic: Transcription

Key Concept	Notes	Words to Look Up	Questions Posed by Professor
(insert key concept here)	(take point-form notes here)	(keep a list of words you need to look up here)	(write down questions that the professor poses to the class)
(insert next key concept here)	...	...	...

Other students find it works to divide each page in two, with notes from the instructor on one side and thoughts/questions that occur to them on the other.

- If the professor incorporates active learning exercises in class, then participate with gusto. Remember that active learning exercises can result in better student performance and greater student enjoyment of the lecture (Haak et al., 2011; Minhas, Ghosh, & Swanzy, 2012; Ueckert & Gess-Newsome, 2007).

- If your professor uses clickers (personal response systems) in class, then obey the clicker rules. If you are not supposed to discuss a question with your neighbour, then don't. Focus on answering it yourself, and then make note of which questions you didn't get right. If you are expected to discuss each question, then stay focused on the question rather than on your social life.

 Go to **science3.nelson.com** for examples of different ways to take notes in class.

 DANGER ZONE

Many of my students try to write down every single thing that I say in class. Do not fall into the trap of having your notes be a catch-all. Think about what you write before you write it.

7.6 STUDY STRATEGIES: RIGHT AFTER EVERY CLASS

After class (either immediately after, or a few hours after), go over your lecture notes and fill in any gaps from memory, enhancing your notes using either your textbook or other resources suggested in class. This forces you to both recall and process information, which will help cement the information in your mind so that you can recall it later on. If you don't understand something from the lecture, then now is the time to see your TA or professor during office hours (see Box 7.3) to get it clarified. Do not wait to do this until the end of term.

BOX 7.3 Making the Most of Professor Office Hours

- Go early during the scheduled office hours. For example, if your professor has office hours from 1 p.m. to 3 p.m., then go at 1:30 pm. Do not wait until 2:55 p.m. Professors often have meetings or other classes to teach immediately after office hours, so they won't be able to give you very much time if you show up at the end of the office hour time block.

- Introduce yourself and say which class you are in. Professors often teach several classes in one term, and they may not recognize you from a particular class.

- Highlight where in your notes things aren't clear if you have a specific question.

- Be sure to bring your assignment with you if you have a question about it.

- Don't assume your professor remembers everything in perfect detail from a given lecture. So, don't say, "the lecture three days ago." Instead, state the name of the lecture and have your notes out and be ready to discuss course concepts.

- Be prepared to come back if time runs out.

- If you particularly liked something the professor did in class (such as an example or activity), then be sure to tell her or him. Professors usually enjoy getting feedback from their students (especially positive feedback!).

After you have enhanced your notes, it is time to go back to the textbook. A good study strategy at this stage is to take the subheadings that are in the textbook and turn them into questions. Can you answer them in your own words based on what you have learned so far?

At this stage of your studying, make use of some of the resources your course instructor provides to enhance your learning. For example, some course instructors set up formal study groups led by students who have taken the course in previous years. Some universities call these groups **facilitated study groups**, or **facilitated learning groups**. This peer-assisted learning approach has been associated with increased learning gains and increased academic performance (Ning & Downing, 2010; Preszler, 2009). Note that the people who lead these groups also often claim to benefit. Many of my students sign up to lead study groups to give them an advantage in their later courses.

7.7 STUDY STRATEGIES: STUDYING THROUGHOUT TERM, AND PREPPING FOR TESTS AND EXAMS

As a university student, you have many different demands on your time. You probably don't have weeks of extra study time, so it is very important that you optimize your study strategy to get the biggest learning gains using the time you do have. Let's talk first about how you will schedule your study subjects.

How Will You Schedule What to Study?

When you prepare for your biology, chemistry, or physics exam, how will you schedule your time? Will you study biology for 1 week straight, 8 hours a day, before the exam? You might be tempted to focus all your efforts on just the upcoming exam, but in fact research shows that it is important to take breaks in your studying. If Group A studies for 3 hours on Monday, and Group B studies for 1 hour on Monday, 1 hour on Tuesday, and 1 hour on Wednesday, research indicates that Group B will likely do significantly better. This is called "spacing," and the term *space* refers to inserting chunks of time between intervals of studying (Bjork & Allen 1970; Cepeda et al., 2006; Kornell & Bjork, 2008).

You probably have lots of different courses to study for and many demands on your time. How will you go about studying for all of your courses? Most people think it would be best to study biology for one night, then physics for another night, and then chemistry for another night. Research actually shows that it is better to mix the topics rather than focusing on just one topic each night. This mixing process also forces you to break up your studying for particular subjects (see spacing, described above). This process is called "interleaving" (Kornell & Bjork, 2008; Richland et al., 2005; Rohrer & Taylor, 2007).

 STRATEGIES THAT WORK

Schedule your study time so that you study one subject for a few hours, then study a different subject, and then a different subject before returning to the first subject. In cognitive psychology terms, this is called "spacing" and "interleaving," and research shows that the best results come with a randomized order.

Condensing Your Notes

So now that you've set your study schedule, you'll want to condense your study notes, as this has been shown to help in the retention of material. One way to condense your study notes is to turn them into concept maps. A concept map is a graphical representation of a particular concept or concepts. You can create a concept map for each key concept covered in a chapter, and then create a summary concept map showing how all the key concepts are connected. It's important to note here that university professors really want you to *understand the connections* between these concepts, rather than just memorizing the individual concepts. Incorporating concept maps into your study strategy can help you achieve true understanding, rather than simple rote memorization, of the course material and can enhance your critical thinking skills (Lee et al., 2011). Let's go through an example of how to create a concept map.

Let's say you're studying the process of DNA replication. To create a concept map of this process, write down a list of all the key terms involved in this process. Next, write down linking words that can explain how these key terms are related to each other. For example, the linking word *unwinds* could connect *helicase* to *DNA double helix*. See Figure 7.1 for a complete list of terms associated with DNA replication, and a concept map created from these terms.

 Go to **science3.nelson.com** for examples of concept maps and printable concept map worksheets that you can use to study.

Practise Applying What You Know

If your science course involves solving problems, then be sure to incorporate doing practice problems into your study strategy. You can get practice problem sets from the end of textbook chapters, the website that accompanies your textbook, and past exams. Also, ask your professor to recommend other places to go to access sample problems.

Do Practice Tests

Many students incorporate practice tests into their study strategy to find out if they are prepared for the upcoming test. However, tests are actually much more useful than just as a

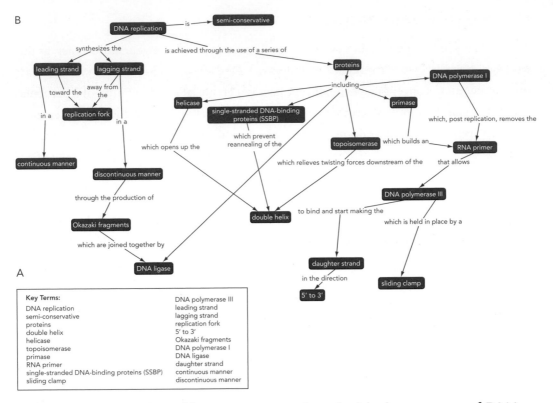

Figure 7.1 (a) A list of key terms associated with the process of DNA replication. (b) A concept map created based on the key terms in (a).

"check your understanding" tool. The process of sitting down and doing a practice test will actually help you to learn the material (Larsen et al., 2009; Little et al., 2011; McDaniel et al., 2007; Roediger et al., 2011). This is because the process of having to recall information will make subsequent information recall easier and more successful (Bjork 1975; Halamish & Bjork, 2011; Hartwig & Dunlosky, 2012). In fact, if Group A studied for 3 hours, and Group B studied for 2 hours and then did a 1 hour test, Group B would likely outperform Group A on a test next week (Roediger & Karpicke, 2006).

Instead of repeatedly studying the same material, be sure to do practice tests and practice exams. Note, however, that for this to be effective you need to make it as similar to a true testing scenario as possible. (This means no peeking at the answers halfway through a problem to see if you're on the right track.) In cognitive psychology terms, the process of test-taking forces you to retrieve information, which in turn makes subsequent **retrieval** easier. This means that the process of retrieval can improve your memory.

STRATEGIES THAT WORK

Test yourself: Insert tests into your study strategy, and make them as realistic as possible. (No peeking at the answers!)

Another thing to remember when you test yourself is that you need to make sure that you include tests of higher difficulty rather than just doing tests that are easy and that you'll ace with little effort. Research shows that if you incorporate hard tests into your study routine, you'll remember the information better and for a longer period of time (Kornell et al., 2009). You can also use tests to uncover your weaknesses and give you a target to focus on for studying. Also, I want to emphasize that you need to focus on what you don't know rather than what you do know. Some students have a tendency to try to get better at what they are already good at in the hope of acing a question on that topic and making up for a deficiency elsewhere. When you have to answer all the questions in an exam, this is clearly a bad strategy, and even when that isn't the case you are playing a game of probability.

It's also a good idea to practise with the type of questions you're likely to see on the final exam. For example, if the final exam is multiple choice, then be sure to incorporate multiple choice tests into your study routine. If the final exam is long answer or calculations, be sure to practise those skills specifically.

How Do You Think Your Studying Is Going?

It is important that you assess your studying as you go along. Do you think your study strategy is working? Do you think your schedule allows you enough time to study all the material? Can you predict how well you'll do on the final exam? Research has shown that people are usually overconfident in how much they have learned through studying (Koriat & Bjork, 2006; Koriat et al., 2004; Kornell & Bjork, 2009). This is because they may understand something when they read it and think they have learned it, but then won't be able to recall it at a later date. In cognitive psychology terms, this is called **metacognition**, or one's "subjective awareness of

STUDY STRATEGIES: STUDYING THROUGHOUT TERM **257**

one's own knowledge." Note that just the act of reading something rarely leads to true under-standing—whereas active engagement in the information, such as through problem solving, can. As you study, you need to remember that you will forget things over time, but if you test yourself on these things you'll improve your recall of them. As we talked about earlier, incorporate tests into your study strategy and make sure these tests include material that you studied weeks earlier, rather than only material you just finished studying (Koriat & Bjork, 2006; Koriat et al., 2004).

Writing the Real Test

When it comes time to writing the real test or exam, you should bear several things in mind:

- Make sure to get enough sleep the night before. (Some exercise is good, too.) As you learned earlier, the best study strategy involves spacing out the material and then testing yourself on it. You don't have time to do this in one night, so don't study into the wee hours of the morning. Make sure to get a good night's sleep! Being sleep-deprived when you write the test can often lead to careless mistakes. An alert mind is your most powerful test-taking tool.

- Make sure the meals you eat before the exam are nutritious and not loaded with sugar. If you have a sugary breakfast (e.g., donuts and a Frappuccino), you'll risk suffering from a sugar crash mid-morning.

- Be sure to check your university's policy concerning mobile phones and academic dishon-esty. Mobile phones are considered to be "unauthorized aids" at many universities and are not allowed in the exam room.

- Some universities require that you place your backpack at the front of the room while you write your test. Unfortunately, many backpacks go missing during the exam, so don't bring any valuables to the exam room. If you happen to leave your cell phone in your backpack, make sure it is OFF—not just on vibrate—because incoming calls can be distracting to everyone. Vibrating phones are NOT silent. Also, please remember to turn off the hourly beeper on your watch.

- When you open your exam paper, the first thing you should do is plan your time RIGHT ON THE PAPER. If the exam is 120 multiple choice questions long and you have 2 hours to do it, turn to question 60 and write "HALFWAY, 1 HOUR," and turn to question 90 and write "3/4 WAY THROUGH, 30 MINUTES REMAINING." This will help you to stay on track as you write. (Note that this technique will only work if you tend to do the questions in order on the test.)

7.8 LABS AND TUTORIALS

As a professor, I often see students rushing through their labs and tutorials with the goal of finishing as fast as possible. The best piece of advice I think I can give you concerning labs and tutorials is to use the time wisely. The more you put into your labs, the more you'll get out of them. Also, the more interaction you have with course material, the better. Especially if this interaction involves active learning and problem solving. If you finish a problem set early in a tutorial, then open your textbook and do some more problems, approaching your TA with questions. All the tips explored earlier also apply to labs and tutorials: You should come to labs and tutorials prepared, be an active participant, and review your notes at the end.

7.9 TIME MANAGEMENT

You may be thinking to yourself, "Oh, sure, these study strategies sound just great, but how am I possibly going to make time for them?" I know that as a university student you have a lot of demands on your time. Perhaps you have a part-time job, are a member of a varsity or community sports team, volunteer, or have family responsibilities or a partner. Whatever the demands on your time, you need to allow yourself enough time to study. You can't rush the study process, and science is a subject you simply cannot cram for. Consider using time management strategies to ensure you have enough time to study.

The first thing you should do is determine whether or not you are a natural procrastinator. Interestingly, research has shown that many perfectionists (and high-achieving students) are natural procrastinators (Flett et al., 1992; Mohamadi, Farghadani, & Shahmodhamadi, 2012). You need to tackle any **procrastination** habits you have RIGHT NOW. People who procrastinate do so for different reasons. Some people are avoiding an uncomfortable task. Others wait to start something out of fear that their work might not be perfect. Whatever the underlying reasons for why you procrastinate, you need to tackle the habit. A very good way to start is by implementing the 2-minute rule (David Allen's rule—I'll introduce you to him in a few paragraphs). This rule means that if something is going to take you less than 2 minutes, then you have to do it RIGHT NOW. Perhaps it means answering an email, returning a phone call, or paying a bill. Don't hesitate. Do it RIGHT NOW. (This does NOT mean that you interrupt a dedicated 1-hour study session to respond to an email. It means that in your free time you don't put off small tasks.) A resource that you might find helpful in combating procrastination is the iPhone and iPad app "Unstuck." This app leads you through a series of questions to determine why you aren't moving forward with a particular project or event, and can even identify the underlying reason why you are procrastinating.

 TOOL KIT

App: Unstuck. Use this app to overcome procrastination by identifying the underlying reasons why you procrastinate.

While we're on the subject of apps, let's discuss some time management apps that you may find useful. Omnifocus and Things are two very popular time management apps among my students. If you use apps routinely, then I suggest you download the trial versions and determine which is best for you. Many different time management apps are available, so you'll probably be able to find one that is a good fit for your needs. The Omnifocus App is based on the work of David Allen. He is the author of the book *Getting Things Done* and specializes in making people productive at what they do—in short, helping people "get things done." The 2-minute rule described a few paragraphs ago comes from his book. A key tenet of his book is that you need a strategy that works for you to organize everything that you need to do. It could be a paper-based strategy like a notebook or an agenda, or it could be an app-based strategy like David Allen's Omnifocus app.

 TOOL KIT

Some apps that are popular to use for time management include Omnifocus, Things, and Wunderlist.

 Go to **science3.nelson.com** for expanded lists of apps, websites, and web-based resources that you can use to improve your time management skills and combat procrastination.

7.10 RECAP: KEY STUDY STRATEGIES TO KEEP IN MIND

1. **Proper preparation is key.** You cannot cram for science courses and retain information or do well on exams. You need to learn new scientific knowledge, and then apply that knowledge through practice, in order to do well. So plan your time carefully and don't procrastinate.

2. **Active learning works.** Incorporating active learning strategies into your study routine (and actively participating in class when your instructor includes them in lectures) leads to greater understanding of the subject material.

3. **Vary your study location.** Rather than studying in the same place every single time, change your study location frequently.

4. **Incorporate spacing and interleaving into your study schedule.** Set your study schedule so that you leave time intervals between studying the same topic, and also alternate studying different subjects during a single study session. (This means study a bit of biology, then a bit of physics, then a bit of chemistry, rather than studying biology straight for 3 days.)

5. **Test yourself.** Check your understanding as you go along, and routinely test yourself. Remember that tests increase learning and retrieval of information. Tests should not be the beginning or end of your study strategy—they should form the middle, wherein you test what you've been studying, and then you take time to learn from your mistakes on the test.

6. **Don't be afraid of failure.** Listen to feedback from your professor and TA, and apply that feedback on your next assignment or test.

7. **Strive for true understanding, rather than just grades.** Remember that what you learn in one course will often be needed as the foundation for a subsequent course. Focus on learning the material rather than the grade you're going to get.

8. **Tackle procrastination.** You need to study effectively and efficiently, which means that you don't have any time for procrastination. Figure out why you procrastinate, and use some of the resources suggested above to improve your time management skills.

7.11 CONCLUSION

Science is a subject that is simply impossible to cram for. You need to study your course topics throughout the term to develop understanding of the content and to become proficient with the skills involved in the subject. The strategies we've explored in this chapter will help you beyond the realm of your science courses—you can apply them to other courses, to learning new sports, and to many other aspects of your life. It's never too late to adjust your study strategy, and you can incorporate the techniques we've talked about above even mid-course. The best time to adjust your study strategy is RIGHT NOW!

FURTHER READING

Allen, D. (2002). *Getting Things Done.* New York, NY: Penguin Books, p. 288.

Assalis, L. A., Giavarotti, L., Sato, S. N., Barros, N. M. T., Junqueira, V. B. C., & Fonseca, F. L. A. (2012). Integration of basic sciences in health's courses. *Biochemistry and Molecular Biology Education, 40,* 204–208.

Bjork, R. A. (1975). Retrieval as a memory modifier. In R. Solso (Ed.), *Information Processing and Cognition: The Loyola Symposium* (pp. 123–144). Hillsdale, NJ: Lawrence Erlbaum Associates.

Bjork, R. A. (1994). Memory and metamemory considerations in the training of human beings. In J. Metcalfe & A. Shimamura (Eds.), *Metacognition: Knowing about Knowing* (pp. 185–205). Cambridge, MA: MIT Press.

Bjork, R. A., & Allen, T. W. (1970). The spacing effect: Consolidation or differential encoding? *Journal of Verbal Learning and Verbal Behavior, 9,* 567–572.

Cepeda, N. J., Pashler, H., Vul, E., Wixted, J. T., & Rohrer, D. (2006). Distributed practice in verbal recall tasks: A review and quantitative synthesis. *Psychological Bulletin, 132,* 354–380.

Flett, G. L., Blankstein, K. R., Hewitt, P. L., & Koledin, S. (1992). Components of perfectionism and procrastination in college students. *Social Behavior and Personality, 20,* 85–94.

Gardner, J., & Belland, B. R. (2012). A conceptual framework for organizing active learning experiences in biology instruction. *Journal of Science Education and Technology, 21,* 465–475.

Goldstein, J., & Flynn D. F. B. (2011). Integrating active learning and quantitative skills into undergraduate introductory biology curricula. *American Biology Teacher, 73,* 454–461.

Haak, D. C., HilleRisLambers, J., Pitre, E., & Freeman, S. (2011). Increased structure and active learning reduce the achievement gap in introductory biology. *Science, 332,* 1213–1216.

Halamish, V., & Bjork, R. A. (2011). When does testing enhance retention? A distribution-based interpretation of retrieval as a memory modifier. *Journal of Experimental Psychology: Learning, Memory, and Cognition, 37,* 801–812.

Hartwig, M. K., & Dunlosky, J. (2012). Study strategies of college students: Are self-testing and scheduling related to achievement? *Psychonomic Bulletin and Review, 19*, 126–134.

Hays, M. J., Kornell, N., & Bjork, R. A. (2012). When and why a failed test potentiates the effectiveness of subsequent study. *Journal of Experimental Psychology: Learning, Memory, & Cognition.*

Koriat, A., & Bjork, R. A. (2006). Illusions of competence during study can be remedied by manipulations that enhance learners' sensitivity to retrieval conditions at test. *Memory & Cognition, 34*, 959–972.

Koriat, A., Bjork, R. A., Sheffer, L., & Bar, S. K. (2004). Predicting one's own forgetting: The role of experience-based and theory-based processes. *Journal of Experimental Psychology: General, 133*, 643–656.

Kornell, N., & Bjork, R. A. (2008). Learning concepts and categories: Is spacing the "enemy of induction"? *Psychological Science, 19*, 585–592.

Kornell, N., & Bjork, R. A. (2009). A stability bias in human memory: Overestimating remembering and underestimating learning. *Journal of Experimental Psychology: General, 138*, 449–468.

Kornell, N., Hays, M. J., & Bjork, R. A. (2009). Unsuccessful retrieval attempts enhance subsequent learning. *Journal of Experimental Psychology: Learning, Memory, & Cognition, 35*, 989–998.

Larsen, D. P., Butler, A. C., & Roediger, III, H. L. (2009). Repeated testing improves long-term retention relative to repeated study: A randomised controlled trial. *Medical Education, 43*, 1174.

Lee, R. M. K. W., & Chiu, Y. K. (1997). The use of problem-based learning in medical education. *Journal of Medical Education, 1*, 149–157.

Lee, W., Chiang, C.-H., Liao, I.-C., Lee, M.-L., Chen, S.-L., & Liang, T. (2012). The longitudinal effect of concept map teaching on critical thinking of nursing students. *Nurse Education Today* (Epub Jul 13).

Little, E. L., Storm, B. C., & Bjork E. L. (2011). The costs and benefits of testing text materials. *Memory, 19*, 346–359.

McDaniel, M. A., Roediger, H. L., & McDermott, K. B. (2007). Generalizing test-enhanced learning from the laboratory to the classroom. *Psychonomic Bulletin & Review, 14*, 200–206.

Michael, J. (2006). Where's the evidence that active learning works? *American Journal of Physiology—Advances in Physiology Education, 30*, 159–167.

Minhas, P. S., Ghosh, A., & Swanzy, L. (2012). The effects of passive and active learning on student preference and performance in an undergraduate basic science course. *Anatomical Sciences Education, 5*, 200–207.

Mohamadi, F. S., Farghadani, E., & Shahmodhamadi, Z. (2012). Individual factors antecedents of academic procrastination: The role of perfectionism components and motivational beliefs in predicting of students procrastination. *European Journal of Social Sciences, 30*, 330–338.

Moore, R., Jensen, M., Hatch, J., Duranczyk, I., Staats, S., & Koch, L. (2003). Showing up: The importance of class attendance for academic success in introductory science courses. *The American Biology Teacher, 34*, 325–329.

Ning, H. K., & Downing, K. (2010). The impact of supplemental instruction on learning competence and academic performance. *Studies in Higher Education, 35,* 921–939.

Newman-Ford, L. E., Fitzgibbon, K., Lloyd, S., & Thomas, S. L. (2008) A large-scale investigation into the relationship between attendance and attainment: A study using an innovative, electronic attendance monitoring system. *Studies in Higher Education, 33,* 699–717.

Obeidat, S., Bashir, A., & Jadayil, W. A. (2012). The importance of class attendance and cumulative GPA for academic success in industrial engineering classes. *International Journal of Social and Human Sciences, 6,* 139–142.

Preszler, R. W. (2009). Replacing lecture with peer-led workshops improves student learning. *CBE Life Sciences Education, 8,* 182.

Richland, L. E., Bjork, R. A., Finley, J. R., & Linn, M. C. (2005). Linking cognitive science to education: Generation and interleaving effects. In B. G. Bara, L. Barsalou, & M. Bucciarelli (Eds.), *Proceedings of the Twenty-Seventh Annual Conference of the Cognitive Science Society.* Mahwah, NJ: Lawrence Erlbaum.

Roediger, H. L., Agarwal, P. K., McDaniel, M. A., & McDermott, K. B. (2011). Test-enhanced learning in the classroom: Long-term improvements from quizzing. *Journal of Experimental Psychology: Applied, 17,* 382–395.

Roediger, H. L., & Karpicke, J. D. (2006). Test-enhanced learning: Taking memory tests improves long-term retention. *Psychological Science, 17,* 249–255.

Rohrer, D., & Taylor, K. (2007). The shuffling of mathematics problems improves learning. *Instructional Science, 35,* 481–498.

Tomanek, D., & Montplaisir, L. (2004). Students' studying and approaches to learning in introductory biology. *Cell Biology Education, 3,* 253–262.

Ueckert, C., & Gess-Newsome, J. (2007). Active learning in the college science classroom. In Mintzes J. J., & Leonard, W. H. (Eds.), *Handbook of College Science Teaching: Theory, Research, and Practice* (pp. 147–154). Arlington, VA: NSTA Press.

Science and Communication

"A scrupulous writer, in every sentence that he writes, will ask himself at least four questions, thus: 1. What am I trying to say? 2. What words will express it? 3. What image or idiom will make it clearer? 4. Is this image fresh enough to have an effect?"

GEORGE ORWELL, *Politics and the English Language*

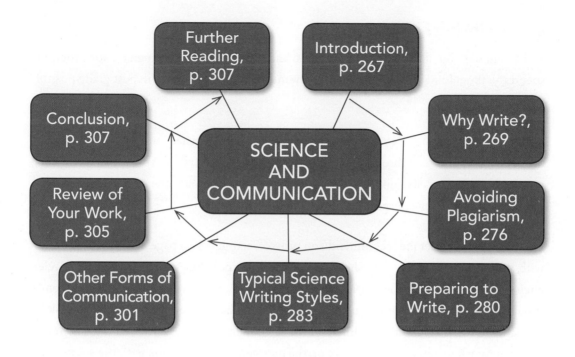

The diagram shows "SCIENCE AND COMMUNICATION" at the center connected to:
- Further Reading, p. 307
- Introduction, p. 267
- Conclusion, p. 307
- Why Write?, p. 269
- Review of Your Work, p. 305
- Avoiding Plagiarism, p. 276
- Other Forms of Communication, p. 301
- Typical Science Writing Styles, p. 283
- Preparing to Write, p. 280

8.1 INTRODUCTION

When communicating, you have to think of your audience. If you're reading this chapter, I figure you're likely a student or an instructor. My words here can serve the instructor by writing simply and clearly for the student, and the instructor will appreciate my writing this chapter for *the student.* So, if you're a student, it's likely you're either preparing or panicking.

Are You Preparing?

You're prudent and you read up before you embark on an assignment. You properly prepare yourself by analyzing the task of writing effectively, and you want to know how to do this well. You're hoping for some tips in this chapter to be efficient and get a good grade.

Or Panicking?

You found out you have an assignment to do and the deadline is coming up soon. You can't spend the time doing this perfectly, so you want quick advice to get the best grade possible under time pressure.

So here's how I'll try to serve you regardless of which category you fall into. If you are preparing, take your time and read through this chapter thoroughly. I'll try to explain why good communication in science is important, and how both the author and the audience use science writing for their specific purposes.

If you're panicking, let me recommend you turn to Section 8.5 on page 283 and do your best with it! You can access the Tip Sheets and exemplar papers for APA, CSE, and other referencing styles online from the *Science*[3] website at science3.nelson.com. Two important points, though: If you're under a time crunch, you're in a population that is at much higher risk for plagiarism or data falsification, both of which are "deadly sins" in science! And please revisit this chapter when you can concentrate on the message; if you continue with the sciences (or dare I say *any* field at college or university), you'll do much better if you take the communication part seriously and learn to do it well. In fact, you're more likely to be hired based on your communication skills than on the experiments you perform: Your opportunities will expand enormously with good writing skills. Your funding (money that organizations will give you to conduct research) will depend on the strength of the proposals you submit as well as your publishing record. If you write well, your work will be better understood and you'll improve your reputation.

Communication comes in many forms. This chapter will deal mostly with written communication, such as **lab reports** or formal essays, but you might get to do an oral presentation or prepare a **poster** at some point. We'll look at this in more detail at the end of the chapter. The preparation you do during research, the need to be cautious about plagiarism, and the act of organizing your thoughts coherently all play a role in science communication. I'll call it "writing" for simplicity, but keep in mind that there are lots of ways to communicate effectively!

The chemistry professors at University of California, Los Angeles, were concerned that their students, although excellent chemists, would not be able to maximize their opportunities because their professional writing skills needed development. For this reason, they created a program called Calibrated Peer Review (CPR), which gave students an opportunity not only to write but also to review and edit each others' work. The CPR exercise gives practical experience in something science professionals do a lot of:

"What do scientists do? Research begins with proposals. Scientists write research proposals and review peer proposals. Scientists do research and write and peer review research manuscripts. Peer review has a prominent role in the progress of science."

CHAPMAN AND FIORE (2001)

8.2 WHY WRITE?

You may be thinking, "This is science, not English! Why should I worry about this?" In this section, I'll try to answer that question.

The Importance of Communication in Science

Science is a process, not a set of facts to memorize. Scientists have to read about current developments in their field. In some cases, the history—where the discipline has been and how it has evolved—is important. Clear, meaningful documentation helps others understand your work, and they will appreciate any efforts you make to enable them to understand what you did, how you did it, and the reasoning behind your work. When you read the work of others and build on it with your own contributions, you are furthering science. Science is self-correcting and subject to change: All participants are morphing our understanding of the physical world through discussions, critiques, and counterarguments. Communication, most often through writing, drives science, and research that isn't communicated is lost.

The Writer's Goals

As a student, you tend to write for others (such as completing an assignment a professor has given you). As you do more writing, though, you'll see that there's value in the act itself. Writing is thinking. If you take care to write things simply but completely, you'll clarify what you know. When you create a sentence, when you reorganize paragraphs, when you delete, edit, and add text, you make decisions about the content and reinforce your understanding. This will help you refine your thinking in the field, and you'll become a more competent scientist. Even if nobody reads your work, you benefit from creating opportunities to exercise disciplined writing.

Others will probably read what you write, though, and that's the more obvious reason to write well. As a student, your work will be judged by educators. Your job should not be to try to please your graders. Instead, use writing exercises as a chance to develop your communication skills, and treat the grade and any comments you get—both positive and negative—as data for improvement. Eventually you might be publishing papers or writing a grant proposal to get funding. Just as the only way to get better at a sport is to practise that sport, the only way to get better at writing is by writing. Any book—including this one—can only make suggestions that you might employ. Simply reading this chapter won't make you a better writer. Only the exercise you get by writing and revising will help you to improve. Everyone needs to start somewhere, and even producing an initial poor paper is worthwhile, as it means you are on your way to becoming a better writer.

❓ Did you know?

In his book *Outliers: The Story of Success*, Malcolm Gladwell postulated the "10,000 hour rule" and gave examples of hockey players and others who became successful at something. That's obviously a lot of time and may undercut the effects of natural talent. However, you might keep this idea in mind when you make a first draft and revise it several times!

The curse and blessing of writing is that there's no single "best" way to do it, and there are lots of opinions about what good writing entails. Clarity is prized by everyone. Many students in the sciences think that their writing should be objective and complex. Papers are often characterized by an utter absence of first-person words (*I, me, we*); long run-on sentences; and unnecessary polysyllabic words. Indeed, some educators discourage "personalizing" lab reports, but in reality, most people find the passive voice hard to read and boring. If you're doing an assignment for an instructor, you might consider asking for advice regarding his or her views about including a **first-person perspective**.

The Reader's Goals

For your audience—the readers—your writing serves to educate. Your readers want to know something. If you're a professional in the field, they want to understand the topic you're writing about. If you're a student, your audience wants to find out if YOU understand the topic. In either case, the advice of Dr. Jan Pechenik (2013) is "write to illuminate, not to impress." Don't try to get fancy. Simple, clear sentences work better than something long, convoluted, and full of technical words. By all means use appropriate terminology whenever possible, but do so in a way that can help define terms for your reader, or model them by using them in an appropriate context.

Regardless of your approach, you'll want to leave lots of time for **revisions** and modifications. You should always write to provide the most clear and interesting description of your topic. If it's a lab report, you'll want to make it easy to understand. Try to do the work for your readers by presenting your data cleanly and in ways that make your lines of logic and your results stand out. Don't jump around with your reasoning: Make sure you collect your points, discuss them, and move on to something else (an **outline** is an excellent tool to organize your ideas!). Articulate *why* you are doing the work. Explain what's known about the field in the **introduction**, and relate that to your research question. Your methods should be clear enough that others could replicate your work. Results should be easy to understand, and you should use text to support your tables and figures so that your audience knows how you arrived at your conclusions. Finally, your **conclusion** ("the fun part," as a colleague once referred to it) gives you a chance to speculate about your hypothesis and where you might go with your results. Did you confirm something you already suspected? Perhaps your ideas didn't hold up in the circumstances you performed the work under. Feel free to speculate about what the results might mean, how they could be applied to other cases, or other studies that could extend your knowledge. In any case, you made a contribution to science that should be shared clearly.

In Box 8.1, I list the top five reasons you should care about writing effectively. Following the box, I expand on each reason.

BOX 8.1   Top Five Reasons You Should Care about Writing Effectively

1. Let's be frank: You want a good mark.
2. Good writing skills are beneficial in ALL subject areas.
3. Writing is thinking.
4. Revisions can help you think of new methods and concepts.
5. The currency of science is your reputation: Good communicators get good opportunities (this includes jobs!).

REASON 1: YOU WANT A GOOD MARK. I remember my time as an undergraduate. We had weekly lab reports, and I remember resenting the time it took for me to put together each one. I kept thinking that the instructors didn't understand that I could make better use of my time by studying for my exams. What my instructors knew and I didn't was that performance on exams and the ability to organize my lab work into a coherent story were separate but equally important skills. In addition, by organizing your thoughts and getting them down clearly, you will be much more likely to remember them on exams and when you do future assignments.

You become a better writer only by writing, and your first pieces take a lot of time and effort and perhaps won't earn exceptionally high scores. Take it from me and from all the colleagues I speak to about this: The more you write, the easier it becomes, and the higher your grades will be because your quality will improve. Advancement in your career will depend much more on your communication skills than it will on the scores you get on exams. Sure, the grades matter for opportunities such as scholarships, grad school, and other professional programs, but don't put all your eggs in the quiz and exam basket! Refine your writing skills.

REASON 2: GOOD WRITING SKILLS ARE BENEFICIAL IN ALL SUBJECT AREAS. I've had students come back to me saying how they changed their opinions about some of the biology exercises we had them do as undergraduates. One story was that the student really hated the formal writing, but while writing a piece to try to qualify for a professional school, all the advice that he resented about writing came back to him, and after he applied it he felt like he'd done a good job (and indeed, got an offer). The ability to organize your thoughts and present them interestingly, clearly, and simply doesn't just apply to communicating science. You can use this skill in the arts, professional schools, and even in letters to politicians and letters to the editor of a newspaper. A good communicator gets more opportunities than someone who is poor at getting her or his ideas across.

REASON 3: WRITING IS THINKING. Students often say they understand something and then, when asked to explain it to others, describe lots of good aspects of the concept but leave out parts and gloss over tricky areas with "you know what I mean" or "I understand this, but can't completely describe it right now." I'd say that the student in fact has an **inkling**, which is something you have on the way to **understanding**. Understanding is what we wish to achieve about an idea: the ability to comprehend the nature or essence of something. If you understand something, you can see how it links to other related things. For example, knowing that an atom forms a bond with another atom demonstrates that you have an inkling about how atoms work. A proper understanding of atoms, though, might include such things as knowing how the numbers of protons in the nucleus affect the bonds that the valence electrons participate in; grasping the wave/particle duality of electrons, and how they're influenced by the nuclear particles; and being able to articulate the *kinds* of bonds that occur between atoms, depending on their orbital shapes and the electronegativity of partner atoms. Actually, understanding how atoms interact at a variety of levels gives you more opportunities to properly predict how a molecule will behave under a number of circumstances.

When you write, you organize your ideas and select the correct words to get the ideas across. Once you've committed your thoughts to the page, you can review them to see if they make sense to you. Writing things down doesn't mean you're done with them! You can always revise your work to make it better and, in the process, refine how you think about them. In fact, you would benefit greatly even if you simply took the time to put your thoughts on paper and then tore it up. Writing is a very active process, and you refine the neuronal connections in your brain as you perform it (Leamnson, 2002). These connections convert short-term memory into long-term storage, making you a better scientist.

REASON 4: REVISIONS CAN HELP YOU THINK. When you get a good idea and want to get it down on paper (for the reasons in the above point, for example!), you might start asking yourself about the details. Why did you use distilled water in your buffer? Why did you boil it for 15 minutes and not 20, 30, or even 16? Would the effect you noticed in your organism hold up for a different species of the same genus? If you think of your audience as you write, you will likely anticipate some questions that will form as they read your work.

Writing down your steps and considering how all the parts relate to the whole is a great way to form a more complete understanding. You might become inspired to extend your work or conceive of other things to examine in the laboratory. Since science is a very fluid process with few boundaries, it's important to draw inspiration wherever you can. As you do with art, you are exploring areas that are compelling because they're unclear and because there's an innate sense of curiosity in all of us that make us explore new things.

REASON 5: THE CURRENCY OF SCIENCE IS YOUR REPUTATION: GOOD COMMU- NICATORS GET GOOD OPPORTUNITIES. The general public often confuse scientists with technicians. Technicians are well-educated people who have learned to use equipment and do precise manipulations that often require skill born of practice and refinement. They need to be able to do the same thing over and over and arrive at the same result. Science, however, involves working in a region of ignorance. Science is done on things we don't understand. Because science needs to work in the blurry areas at the edge of the unknown, it's critical that the scientist be able to communicate exactly what's already known in the field, what the questions are, and how his or her tests of the unknown are supposed to reveal new things.

In some cases, like the multibillion-dollar Large Hadron Collider in Europe, the funding was acquired by many physicists being able to articulate (1) what was known about the universe and the nature of matter, (2) why it is important to extend that understanding, and (3) how extending humanity's comprehension of the material of the universe can lead to a better quality of life for everyone. Because of successful communicators in the physics community, a multibillion-dollar facility that literally spans several countries was created. It's unlikely you'll be competing for such enormous amounts of money, but funds are always tight, and they are awarded to those who can best make their case.

And it's not just money that scientists need. Science has a strong component of dissemination. This can be through talks (also known as "**presenting a paper**" because it's often accompanied by a written piece that is published after the **conference** or gathering), posters (visual displays that summarize a small portion of a laboratory's work), and of course journal articles and other written pieces. In order to be invited to contribute to any of these, you will probably have to submit an application to have your work accepted. Conference boards understand that it's important to give up-and-coming scientists opportunities to present, and they will look favourably on well-written **abstracts** (summaries of the presentation). **Keynote** speakers and **panel** members involve women and men who have made their mark in the field. They have proven that they can communicate their ideas clearly, and they have been successful with their well-written grant applications. They are given more opportunities to present their work and in many cases have to turn down some of these opportunities. Your ability to write to illuminate will inevitably impress people in your field, which will result in better opportunities as your reputation spreads. A brilliant scientist will not fare particularly well if he or she cannot explain his or her work and how it relates to the larger context.

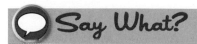 **Say What?**

ABSTRACT: This artsy-sounding word does not necessarily mean something that's not concrete or without a physical basis. In science, it basically means "summary." One often thinks of a summary as something that comes at the end. However, an abstract serves to stand in the place of the complete work and is therefore used to describe a paper (online, for example, or in applying to have it accepted for a formal presentation). It's usually found right after the title of the paper for this reason: the title serves as the first flag for attention from a reader so she or he can decide whether to read the rest of the work. The abstract is quick to read and digest and is the second opportunity for the reader to decide whether to devote time and attention to the whole article. Time is precious, and a whole protocol for quick communication of ideas has been established. The abstract is an important component that allows scientists to keep informed of developments in their field without becoming buried in an avalanche of information.

8.3 AVOIDING PLAGIARISM

Why Is This Important?

Now comes a critical part. Scientists create knowledge. We work in a strange environment where we generate ideas that we want to share, and—indeed—we make our living from sharing. **Attribution** is critical to provide proper **credit** for the originator of an idea, because we require the same courtesy when we come up with something novel or insightful. How would you feel if someone else got full credit for something you created and published? In order to think bigger thoughts, we need to read up on the work of others and integrate their ideas to form new ones. In a letter to his contemporary, Robert Hooke, Sir Isaac Newton borrowed a phrase from Bernard of Chartres: "If I have seen a little further it is by standing on the shoulders of Giants." No scientist works in isolation: He or she succeeds by understanding the work of others and extending or adapting it with his or her personal insight.

What Is Plagiarism?

You might be asking, "what constitutes **plagiarism**?" This chapter cannot set policy for institutions or specific instructors, and so you must ask your instructors about their definition of plagiarism. More progressive universities take an educational approach to plagiarism. Instructors who find ideas that did not originate from the student and are not attributed to their source might report the infraction to an Office of Student Conduct, which then takes the time to explain to the student what plagiarism is and give special instructions about how to avoid it. In addition to the education, there might also be some form of score penalty, which is important to dissuade students from "testing the waters" by throwing in the work of others to add bulk to their writing. Plagiarism has been described as "vicious" by Miah and colleagues (2011), where they present it from an interesting viewpoint. By taking credit for the work of someone else—especially a peer—you are depriving her or him of credit for the idea and taking it yourself. This is theft. You are also guilty of counterfeiting: You are passing yourself off as the creator of something that was made by someone else.

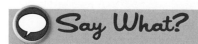

Say What?

PLAGIARISM: The word plagiarism comes from the Latin word *plagiarius*, which literally means "kidnapper." Plagiarism is the presentation of someone else's work or ideas as if they are your own. This can be intentional (such as handing in someone else's assignment or using someone else's sentences to receive credit for work you did not do) or accidental (for instance, by forgetting to put a reference to the work's source). Plagiarism need not be specific words in a specific order; you can commit plagiarism by paraphrasing someone else's ideas as if you thought of them yourself. Misuse of others' data or words is also sometimes referred to as plagiarism.

Fair Use of the Work of Others Isn't Plagiarism

Students are sometimes concerned about submitting writing that contains a lot of work by other scientists. Provided proper attribution is granted—that is, the student clearly distinguishes ideas that are borrowed from others from his or her own—then it's a very good and, in fact, inevitable practice. It's important to rephrase things and provide your own interpretations, though. In a way, this is like the common practice of creating new works from mashups on YouTube or the like! The combinations are intellectually stimulating and provide new perspectives. The work that is borrowed must give proper respect to the originators, though, and even if you **paraphrase** an idea, you need to give credit to the originator of the idea. By giving credit (which costs you nothing, unlike the reuse of a melody or photograph), you are helping others see the contribution that the idea's original author made to the field.

Plagiarism also occurs when someone misrepresents the work of others to suit her or his own agenda. Sometimes **references** are chosen based on the title alone and not actually read and considered. In the course of writing this chapter, I did some research on plagiarism and—in searching for data about plagiarism—actually found many examples of misattribution in the literature itself! Data from the United Kingdom were presented as if they came from U.S. students, and some articles were not identifiable possibly because of misspellings in authors' names, violating the responsibility the author has to his or her readers—for readers to be able to find and read the source material to see if the author's interpretation is similar to theirs.

Plagiarism is not rare in academic settings. Many surveys show that half or more of students engage in plagiarism either willingly or accidentally (Miah et al., 2011), although other studies show that the plagiarism rate is much lower (Walker, 2010). Regardless of the frequency, plagiarism is always wrong, and the consensus among educators is that transgressions require penalties. Interestingly, it is not only students who plagiarize. There have been high-profile stories in the news of ranking academics who misappropriated the work of others (Gillis, 2011) losing their positions as a result. Even political figures are sometimes accused of academic theft, and in some cases there are no sanctions, such as when the then-premier of Alberta, Ralph Klein, submitted portions of a paper without proper **citations** to Athabasca University for a course he was taking in 2004.

Taking Notes

How can you avoid plagiarism? First, *do not cut and paste information from your source into your work.* It's easy to forget to paraphrase and attribute work if you get interrupted. If you paraphrase your notes on the work of others, it's also easy to change it back unintentionally to the original wording "because it sounds better" than your notes! Some students use the strategy of writing their sources verbatim into their notes, or keeping a collection of the original source papers (either digitally or as hard copy) and always **summarizing** and attributing the work immediately after it goes into the **draft** on the word processor. It's common for a student to worry about missing a source and to wonder just to what extent she or he needs to make citations. This is a skill that comes with practice. Sometimes students commit plagiarism accidentally and face sanctions (Teitel, 2011), so keeping good notes is critical not only for avoiding plagiarism in the first place, but also for demonstrating how you tracked your resources in the unlikely situation where your professor needs to follow up on a mistake. As a rule of thumb, you should make the mistake of overattributing work. It's not much work to reorganize your ideas after you've got them down to consolidate your references.

In Box 8.2, I give some tips on **quoting**, paraphrasing, and summarizing.

BOX 8.2   Quoting, Paraphrasing, and Summarizing

Quotations are word-for-word representations of an original work. They must be clearly identified as coming from another source, most often by putting quotation marks around the words that were taken, followed by a citation (an identification of the source). In science, it's usually unnecessary to quote from others. Writing for the humanities deals with subtle uses of words and rhyme and meter. Science should not be subtle: It needs to be clear. You should avoid using quotations.

Paraphrasing is when someone else's words are put into a new form that preserves the original sentence's structure. It, too, needs to have a citation.

Summarizing is a synopsis of what someone else presented. It doesn't use the same structure as the original source. This needs to be cited (see a pattern here?).

☠ DANGER ZONE

Students who are trying to finish their assignments under a time crunch are much more likely to plagiarize someone else's work or to falsify data. Whatever your situation, do not commit either of these "deadly sins" of science. Many instructors will deduct 10 percent of your final mark for an assignment that is late by one day. It's better to take a mark deduction than it is to falsify your work. And it is of course better still to avoid rushing at the last minute to finish your assignment. So start early and try not to leave your work until the last minute.

Another important activity that reduces the instances of plagiarism is *demonstrating good time management*. When I talk to students who have plagiarized material, over half say that they got too busy at school and the deadline was imminent. They used "cut and paste" in order to get the assignment done in time. Of course the penalty for a late assignment is less severe than for fraud, so the strategy turned out to be a very bad idea. Revision of your work is essential, so try to get it done far enough before the deadline that you can take a break and read it with fresh eyes. You should also consider having a classmate—someone you trust won't take your work and submit it as his or her own!—look over your work as an editor. You can strengthen your communication skills by serving as an editor to him or her, too!

8.4 PREPARING TO WRITE

There are lots of ways to communicate science. Sometimes it's an essay, sometimes it's a formal lab report, and sometimes it's a non-written item such as an oral presentation. In all cases, you're best to do your research *before* you get too far along in your actual presentation.

The Wrong Way

It's very common for students to go about this the wrong way. The wrong way is to start writing and then find references to support the ideas you've already come up with. In a lot of cases, you might be able to pull this off: The project might deal with a concept that you were taught in a lecture, so you can just parrot the basic ideas onto a piece of paper and reference the textbook. This isn't science, by the way: Science is a *process* by which we find out information about the world around us; it's not just a list of facts.

Do science labs deal with science? Interestingly, most teaching laboratories don't actually involve science experiments. Many laboratory exercises involve carrying out steps of a process to arrive at a specific, predetermined answer. In many cases, the university laboratory deals with "discovery science," which is a careful and systematic way to collect data regarding the system we're investigating. In Chapter 2, we examined experimental and discovery science in more detail. The data you collect in the teaching laboratory may not extend humankind's knowledge, but at least it's new to you!

If you're doing an actual experiment, you're often exploring a situation where the answer isn't absolutely known beforehand. In this case, doing research before you start writing becomes more important. If you're given a system to study in which you get to select your own variables, it's important to get to the library or online to find information that will help you construct a good hypothesis. It's common for students to go about this the wrong way: They jump in and start gathering data, and only when writing up the results do they try to find references. At this point, you might not find much research that connects your data to the hypothesis, or you might have missed a more interesting and useful question to ask.

Students often approach me after doing the experiment, concerned that they can't find references to fit their data. This betrays the poor decision to "do the work now and get supporting evidence later." Remember, science is not intended to show *why* you're right. If you're looking only for validation, you're susceptible to bias. Scientific investigation should impart new information that extends the blurry edges of what we know. Serious science does not merely confirm concepts we're comfortable with, and poking in references ***ex post facto*** that support what we already know doesn't provide much benefit.

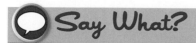

EX POST FACTO: Latin, meaning "after the fact." These are activities that you do after the information is collected or the event has occurred.

The Right Way

If you've been given the opportunity to "do science" in the lab, for example, by designing and carrying out your own experiment, it's very important to do your literature search before you decide on a hypothesis. It's vital to go to the library and find relevant information about your topic. For example, if you're researching the properties of an enzyme, find some papers that describe where the enzyme is normally found, what kinds of processes it catalyzes, and what kind of research has already been done with it. It's not a bad idea to go searching for information after you've roughed out an idea about what you want to investigate. For example, you might have an idea that the enzyme you're studying might work best in a basic environment due to its location in the small intestine. As you find articles, you might become interested in a cofactor that you discover while reading the work of others. You might think of more interesting ways to do your experiment as you gather information.

The Role of the Librarian

Nowadays you don't need to go right to the stacks to find information; you can do this from a computer using Google Scholar or links through your school's library webpage. Most universities have subject librarians who are happy to help you find articles in their field. They are information specialists, and they chose their career because of their interest in finding information and creating search strategies. There should be a help desk somewhere in your library. Go there and explain what you're looking for. You might be directed to someone whose job it is to find information about science. She or he can give you tips on searching and demystify the hunt for good information. It's well worth introducing yourself to the librarians—their job, contrary to popular belief, is NOT to reshelve books or ensure a quiet working environment. At some universities, instructors meet with their subject librarians to give them details about assignments that might require research. This means that they can give you tips for understanding what your assignment requires.

With the crush of other projects in your courses, it's sometimes hard to justify a trip to the library well in advance of an assignment deadline. However, keep in mind that you need to make the trip at some point. Get to know the library early on: It will make your life *easier*. You'll get a greater benefit for the same amount of time if you do your research before doing the experiment. Reading papers related to your topic can be inspiring. You can get ideas that might make your project better.

And speaking of being prepared, be sure to read through your laboratory manual before each lab! Failing to do so can actually be a safety hazard, because details of the chemicals and equipment you might be using will include proper handling tips. Preparing can also save time: If you keep a lab notebook, you should prepare your tables in advance. For example, if you read your lab manual before the lab and see that you'll be taking measurements each minute for 10 minutes, you can create a **table** in which you can enter the data as you collect it. This saves time in the lab, and you'll have a better understanding about the reason you're doing the work in the first place.

To summarize: There are things you need to do before you roll up your sleeves and start banging on the computer keys! In Box 8.3 I remind you of three of these things. Do them *before* you put your masterwork together!

BOX 8.3 **The First Three Steps before Writing**

1. Think about what you're trying to say. Form an outline.
2. Visit the library to do your literature search. You could enlist the help of a content librarian.
3. Find out what form your writing should take. If you're doing an oral presentation, get your thoughts organized on paper. If it's a formal report, find out if it should be APA, CSE, or some other format of writing style.

8.5 TYPICAL SCIENCE WRITING STYLES

To ensure consistency and help their readers get their information clearly, journals have defined writing styles that all contributors must adhere to. Headings, font size, and page layout are described by the journal: Papers that don't conform to their requirements are rejected outright.

The Journal: The Final Arbiter of the Format

Journals draw their guidelines from several basic styles. Chicago, IEEE, MLA (Modern Languages Association), SAA (Society for American Archaeology), and ACS (American Chemical Society) are used for various disciplines, but this chapter will focus on APA and CSE. **APA** stands for the American Psychological Association and is widely used in areas other than psychology communication. Chances are very good that your librarians are familiar with this style and probably have a tipsheet that can help you understand the requirements for writing in the required format. Unfortunately, APA style is in a constant state of change, and it can be hard to keep up to date on the requirements. There are a large number of online resources that help walk you through APA requirements. Due to its widespread use, I've included a little instruction about APA here and linked to resources online. **CSE** is the Council of Science Editors format, which, obviously, is appropriate for writing in the sciences. CSE is not as prevalent at universities but in many ways is similar to APA.

APA and CSE Styles

In both APA and CSE, information is organized into sections, particularly when you're reporting an experiment. The first page is the *title page*. In APA, there's a separate page for the *abstract*, which serves as a **summary** of the entire work. After that is the *introduction,* then *method,* then *results,* then *discussion,* and finally *references*. If necessary, you can add an *appendix*. For CSE, there are similar sections, but—depending on the editor or professor—there is no abstract. If that is the case, there is a *summary* at the end of the paper.

Author-Year and "In-Sequence" Citation

You'll find that many librarians and university-specific online guides focus on referencing. APA and CSE are both designed to make retrieval of sources unambiguous and easy.

In today's world, we have the benefit of an amazing array of sources. They can be in print or live interviews, found online as videos, posted as blogs, and found in all kinds of ephemeral and varied forms. As noted in the section on plagiarism earlier, it's really important to give proper credit to the originators of the ideas you're using.

APA uses an **author-year format**, which means that you embed the authors' names and the year of publication right in your text. CSE supports both author-year and **in-sequence citation**. The latter format is common in chemistry journals and involves superscript numbers that link the references at the end of the paper to the places in the text where the references were used. If you're using author-year format, the references are alphabetized by the lead author's last name. The in-sequence citation method lists the references in the order that they appeared in the paper (see Figure 8.1).

The format of the different levels of headings and the names of some of the sections differ between CSE and APA. One final difference between APA and CSE is that for the author-year formatting, there's a comma between the author's names and the year in APA, but no comma in CSE.

a) This is a sentence using author-year format (Smith, 2012). Jones and Adams (2011) is also used here to provide another example.
b) This is a sentence using in-sequence-citation format.[1] Jones and Adams[2] is also used here to provide another example.

Figure 8.1   Different styles of citation. a) Author-year format has the last name(s) of the author(s) and the year of publication. These appear in the References section in alphabetical order (see the example online). b) In-sequence citation uses superscripts to link to the references in the References section. Note that if you add or remove references, you need to renumber in the order in which these appear in the text, and your References section must be reorganized so that the references are, well, in sequence!

In Table 8.1, I list a few examples of APA reference style. Other examples can be found online.

Table 8.1 Examples of APA Reference Styles for a Number of Source Types. To see an extended version, along with an APA tipsheet, go to the *Science*[3] website.

BOOK				
Author.	(Date).	*Title of book.*	Location:	Publisher.
Kandel, E. R. (2012). *The age of insight: The quest to understand the unconscious in art, mind, and brain, from Vienna 1900 to the present.* New York, NY: Random House.				

PERIODICAL (PRINT OR ONLINE)					
Author.	(Date).	Title of article.	*Title of Periodical,*	*Volume*(Issue),	Page(s).	DOI (if available)
Iezzoni, L. I., & Ogg, M. (2012). Patient's perspective: Hard lessons from a long hospital stay. *American Journal of Nursing, 112*(4), 39–42. http://dx.doi.org/10.1097/01.NAJ.0000413457.53110.3a					

PERIODICAL (NO AUTHOR)				
Title of article.	(Date).	*Title of Periodical,*	*Volume*(Issue),	Page(s).	DOI (if available)
Borderviews: Light plain/plane of light. (2012, March/April/May). *Border Crossings, 31*(1), 18.				

ONLINE/DATABASE ARTICLE (NO DOI)					
Author.	(Date).	Title of article.	*Title of Periodical,*	*Volume*(Issue),	Page(s).	Retrieval Statement
Jennings, H., Nedeljkovic, M., & Moulding, R. (2011). The influence of confidence in memory on checking behaviours. *E-Journal of Applied Psychology, 7*(2), 9–16. Retrieved from http://ojs.lib.swin.edu.au/index.php/ejap/article/view/192/278					

ARTICLE AVAILABLE ONLY IN A DATABASE (NO DOI)					
Author.	(Date).	Title of article.	*Title of Periodical,*	*Volume*(Issue),	Page(s).	Retrieval Statement
Thompson, K. J., Thach, E. C., & Morelli, M. (2010). Implementing ethical leadership: Current challenges and solutions. *Insights to a Changing World Journal,* (4), 107–130. Retrieved from http://www.franklinpublishing.net/insightstoachangingworld.html					

Source: From Haig/Raikes/MacMillan. *Cites and Sources 4E.* © 2014 Nelson Education Ltd. All rights reserved. Reproduced by permission. www.cengage.com/permissions

 TOOL KIT

Tipsheets (CSE and APA) are found online, as are sample papers.

Note that headings are bold, have initial word caps, and are centred in APA. In CSE, the major headings are in caps but not bold and are left-aligned. The structure of the title pages is different between the two styles. APA uses an Abstract, which occurs after the title page, but CSE may have a a summary that is found after the Discussion section and before the References. The experiment is described in the **Methods** section of APA, but it is named **Materials and Methods** in CSE.

Ladies and Gentlemen of the Jury. We Intend to Show...

A well-designed lab report in some ways resembles a court case. If you watch a trial on TV (real or one of the fictional court dramas), you'll notice that the opening statements set up the case. This is how your introduction should work. The research you did before performing the experiment will inform much of your introduction. The introduction gives background information relevant to what you're studying in the experiment. The facts you uncovered during your library research are presented and referenced. Remember that your jury consists of intelligent people, but they may not be as familiar with the system as you are: You are using your lab report to educate them.

During the trial, the police, forensics team, and expert witnesses will outline how the evidence was collected and what it means. This is your Method section. It is not intended to create an expert in the way the research was carried out. It does, however, point to other experts who have more detail about the way the procedure is normally carried out and contains information about how protocols have been modified. With a little extra training or guidance from experts, the members of the jury should be able to do the work themselves.

The results represent the evidence. They should be organized and simply presented. The audience should be able to draw their own conclusions from what's been given, but it's alright for the lawyers to point out trends. Last comes the interpretation: the **discussion**. The legal team shows whether the results support or contradict the theories laid out before. To add credence, the teams might throw in extra analysis with an editorial explanation of how this case affects current law or politics.

It's important to note that a real trial involves skills of persuasion, and the prosecution and defence are paid to hold a particular position. Science needs to be objective. The analogy here is that you need to clearly present what is known and how the data are to be collected and interpreted. You shouldn't do science with an agenda beyond trying to understand the natural world.

If you keep in mind the purpose of your writing as you do research and get actual words down on paper, you might find it easier to select the points you want to present and organize your ideas. You're educating someone else when you write. You have ideas and want to present them clearly to your audience so they can understand why you hold the beliefs you do. You have a sense of the discipline and want to help others understand how you think about it. The story in a paper is a powerful tool to that end.

Unlike a real jury, your audience will largely self-select. This means that your readers will be people who've scanned two pieces of information and decided to dedicate a little time to reading more of your work. These two pieces of information are your title and your abstract. From that point, they are likely to skim through your introduction and jump right to your results or discussion. If these parts catch their interest, then they'll dedicate a little more time to the methods and details you've written. For this reason, you should be careful how you organize your ideas. The reader will have a certain expectation of where to find different types of information.

By understanding the purpose of each part of the lab report, you can better decide what part belongs where, and you can decide on the most appropriate way to write each section. In the next section I list the common sections according to APA style. At the *Science*[3] website at science3.nelson.com, you'll find the same advice, but tailored to CSE style. Be sure to check with your instructor about his or her preferred method of communication. They're similar but—as noted before—each journal (or instructor) has the right to insist on a particular style! One last point that will be revisited later: Although these sections are presented in the order that they will appear in the **manuscript** that you hand in, in reality you will write them in a different order! Don't worry; it'll make sense when you hear the details. Box 8.4, later in this chapter, guides you through a logical order of activities!

 Go to **science3.nelson.com** for examples of a paper written in both APA and CSE styles.

Sections of the Manuscript

For APA, all your text should be double spaced, on 8.5" × 11" paper ("normal size") with 1" margins and use 12-point Times New Roman typeface. Left-justify your text. This means that all unindented lines start exactly at the left margin and the right side of each line is "ragged" (not all lines are of exactly the same length).

Headings should be indicated using a standardized system. The highest level (Level 1) headings are Title, Abstract, Introduction, and so on. These are centred, use boldface, and have the initial letter of the key words capitalized. The heading is on a line of its own (the text that follows appears on the next line). Level 2 headings are the same except that they are left-justified. Level 3 text is marked by letters in bold typeface with all words after the initial word (except for acronyms) in lower case. The same is true for Level 4, which is bold and italicized, and Level 5, which is italicized only. Level 3, 4, and 5 headings are all indented from the left margin and end with a period, and the subsequent text follows immediately.

The section that follows describes what we've been calling a "laboratory report." In some APA guides (such as Angeli et al., 2010), this may be called an "**experimental report**." You'll usually use only Level 1 and 2 headings in an experimental (laboratory) report, and subsequent levels of headings if you are writing a research paper in APA format.

The Laboratory Report (Experimental Report): APA Format

TITLE. In the arts, a title should be clever and sometimes mysterious. It should attract the attention of readers like a lure. In science, however, your audience won't be impressed by a clever title that attracts them if the paper doesn't meet their needs. Science takes time. Whether you're a student or a researcher, you probably want more hours in the day and have a huge "to-do" list of activities that crowd the time you've already carved out for yourself. The title for a science paper needs to be informative and accurate. It's the first cut regarding whether it's useful to the person who scans it.

For this reason, the title should be fairly long and descriptive. Officially, APA style suggests a title of 12 words or less, but this is pretty restrictive. You don't want to waste words, though (time is of the essence!), so you should try to describe in a single sentence what the paper is about. Include the genus and species of organism if it's a biology paper, so a primatologist won't end up pulling a paper that deals with a hormone that causes moulting of a chiton. If you're writing about the bond angles in a novel toxin, name the toxin using appropriate terminology and describe an important feature you discovered.

In APA style, the title page should be a separate page. All pages are numbered, and this is page number 1. In the header, you should define your running head. This is a phrase that captures the title (but is not the complete title) and is typed all in caps. The header of every subsequent page should have the running head left-justified and the page number right-justified.

ABSTRACT. If the title grabs the attention of a reader, she or he will likely next jump to the abstract. Remember, the abstract is a summary that has a little more detail about the larger work. The reader can get a better sense of what was being investigated and the experimental outcome. The abstract should provide the reader with enough information to understand what the overall paper is about. It should be brief (depending on the format, 130 words maximum) and yet cover the purpose, *very* briefly the method, the overall result, and perhaps even the conclusion of the work.

In some cases you might provide a list of keywords. These are provided on a separate line (after the italicized statement *Keywords:*). Although keywords used to be critical for database searching, whole-text searching is making them less necessary.

INTRODUCTION. Undergraduates often feel insecure about the level of detail that should be used for the introduction (in fact, I vividly recall discussions with my classmates where we all felt like we were a bit lost as to how much to put in!). This is very normal, and the cure is simply to practise. While you should be careful not to write in a complex way in an attempt to impress your audience, you should still use appropriate vocabulary in suitable situations. A cool mnemonic is K.I.S.S.—keep it simple and straightforward. Show your audience how to use the words. The introduction does not need to reiterate all the details of your reference papers, but it should tell a story. With long-form writing, there's no single correct way to state your ideas. The important things to keep in mind are relevance and clarity. Guide your reader through the scope of the project (what you are studying), and explain what is already known (why you selected the references). Explain what you're manipulating (the independent variable) and how it's related to what you're measuring (the dependent variable). There should be a clear line of reasoning connecting these items. The purpose of your experiment is to test whether the connection between your independent and dependent variables makes sense based on the logic you outline here. Present your case.

In the arts, students are sometimes encouraged to start the work with a question. This is a good technique to get the ball rolling, and you might think in the sciences that using a good question as the first sentence of the introduction makes sense. However, this tends to undermine the objective tone of the paper. Instead, it is better to let the introduction build up to the question by presenting the logical flow of reasoning that results in a hypothesis.

Structure of the Introduction. One good strategy for how you present your information in the introduction is to think of it as a funnel. Start out broadly, and in your first paragraph explain the overall relevance of your system to your reader. The next paragraph could then zero in on the particular aspect of your field and why it's interesting. You might have a few more paragraphs that outline theories other scientists have constructed and tested regarding the system and explain how your variables will be manipulated and measured. The final paragraph is thus set up for high impact: It will deliver the hypothesis that is the reason for the paper to begin with.

End the Introduction with a Good Hypothesis. While the common way to describe a hypothesis as an educated guess, this description is misleading. A guess implies conjecture and usually lacks enough information to have a very good chance of being correct. A hypothesis is better defined as "a proposed explanation" (see p. 60 in Chapter 3). It's based on the reasoning set forward in the introduction and on the evidence that the supporting references have supplied. While a hypothesis *must* be falsifiable—generating a test that will distinguish whether the reasoning behind it is sound or not—a good hypothesis is based on a thorough understanding of the system and an acquaintance with the work that's already been done. You're not investigating a guess, but rather you're pushing back the areas of ignorance at the fuzzy edges of science.

One way to create a hypothesis is to use an "if–then" construction. After you've defined the problem under investigation in the preceding paragraphs, you can state, for example, "If hormone X is a vasoconstrictor in the parasympathetic nervous system, then application to Y tissue should give result Z"; or "If there's a conjugated bonding system in chemical X, then the NMR spectrum should demonstrate peaks at Y." I hope you can see how the preceding paragraphs set up the logic behind these statements. Another technique that's consistent with this type of hypothesis is "I wonder if…." Depending on your instructor, you might be discouraged from writing in the first person (it's not incorrect, but different scientists have different preferences). A hypothesis is a well-reasoned statement that tests the linkage between two variables. In the natural world of cause and effect, science is our tool for examining these linkages. If you have a logical, clear line of evidence that explains natural phenomena, your hypothesis will state *how* the variables are related and *why* there should be a connection.

You should identify where the information in your introduction came from. Be sure to cite your references clearly. When you think you've done your last revision, go through and correlate all references you cite with your reference list. Remove all items in your list that are not cited, and if you make a citation and don't find the reference in your list, make sure you fix it by deleting the citation or finding the reference!

METHOD. This part lets your reader know how you tested your hypothesis. The assumptions and logic behind your protocol should have been laid out in your introduction. Since you did this already, be sure to present this in the past tense. Depending on your lab instructor's preferences, you might just be asked to reference your laboratory manual. However, an actual formal manuscript will consist of several paragraphs that outline how to perform the experiment. It's acceptable and appropriate to reference other papers that have used the same protocol as you did, but be sure to note any modifications you made to the procedures you cite. Between what you write here and the references you cite, the reader should be able to replicate this experiment. If you're using equipment that might influence the data you collect, you might want to note its make and model. For example, a centrifuge is fairly standard, and except for exceedingly precise protocols, won't affect your results. However, a mass spectrometer needs special calibrations and does some specific data interpretation before displaying the data, so you should note the kind of equipment you used. If the manufacturer of the equipment notes a bias later on, you and other researchers might need to reconsider the findings.

Note that a very common error for this part of a formal laboratory write-up is to create a bullet list of equipment and sample material. Although this is a reasonable thing to put in your laboratory notebook (which is where you should be keeping your research notes), it has no place in a formalized piece of writing. Don't make this section look like a recipe card or a shopping list. Also be aware that you shouldn't put your results here. They belong in the next section. If you find yourself explaining rationale in this area, consider putting that in the introduction. For specialized readings, you should record the make and model of equipment that you use, particularly when reproducibility might be at stake!

Remember: "A place for everything and everything in its place."

RESULTS. Here is where you put your data. You shouldn't put your raw data here (they belong in your laboratory notebook). Instead, process your data. Take averages. Calculate the variance between replicates. Note that some types of data suggest the presentation style outright. For example, if you're looking at the effect of temperature on a physical process, such as conduction of electricity in a metal, then a line graph makes the most sense. In other cases you might choose a table. Different people have different preferences for how the data should be displayed. What all agree on is that you should make the job as simple as possible to interpret, and you must be sure to choose only ONE way to present the data: Don't try to hedge your bets and present the same data in two ways. The reader will be annoyed with you for wasting his or her time! Your presentation style should make trends in the data as obvious as possible. In addition, since you've already done the experiment and these are the data you collected, you need to write this in the past tense (just like in the Method section).

Statistics are the foundation for data. If you run many replicates and find out that one condition consistently gives higher numbers than another, you still can't claim that there's a consistent, reliable effect. Statistics are a mathematical way of assigning certainty to the conclusions you're drawing. For example, if you analyze your data in a particular way, you could say that you're 99 percent confident that the conclusion—that the independent variable affects the dependent variable—is correct. If you see political polls on the news, they often say "this is accurate 19 times out of 20" because of the statistics the pollsters used to analyze the data. They are saying they are 95 percent sure that the poll reflects the views of the general population. It's another way of saying there's a 5 percent chance that the data do not reflect the views should they have taken them from every single person. This isn't the chapter to go into statistics: You should look at Chapter 5 for more information.

For each table or figure, you need to have a caption and an identifying number. The number increases consecutively from 1, and the numbering system is independent between tables and figures (Table 1, Figure 1, Figure 2, Figure 3, Table 2). **Figures** usually include both graphs and photographs. After the identifying number, you need to have a caption: one to three sentences explaining what the figure shows. Symbols need to be explained here, and the sentences of the caption should provide enough detail to explain the figure on its own. Note that you'll also have accompanying text in the results section to help show the trends. The caption can include information such as the number of replicates done, the types of samples used, and information about units used for numerical data. There should be enough information so if the figure is reproduced (as in an oral paper, for example—see below), then it can stand alone. The number and **caption** go ABOVE tables and BELOW graphs (Figure 8.2). See the online supplement for more examples.

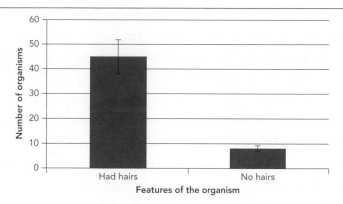

Figure 1   This purely hypothetical bar graph demonstrates how to show data in a bar graph. Note that this informative caption occurs underneath the figure to which it refers. A bar graph is appropriate because the organism either had hairs or did not have hairs. Error bars show variation between different measurements.

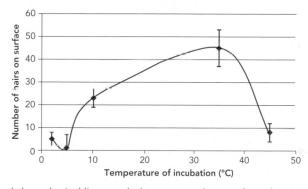

Figure 2   This purely hypothetical line graph demonstrates how to show data in a line graph. Note this other informative caption. A line graph is appropriate because the data vary over a range of temperatures and you can suppose that an intermediate number of hairs will grow if the organism is incubated at 22°C.

Table 1   This purely hypothetical table demonstrates how to show data. Note that this informative caption appears above the data presented, unlike in figures.

Data Point	Value
1	45.3 ± 3.1
2	21.4 ± 9.6
3	15.8 ± 3.3

Figure 8.2   A figure of figures (and a table). Note that all have captions that give enough context to the viewer so that it could stand alone. Your manuscript will refer to the image by name (Figure 1 or Table 1, as appropriate). Captions go underneath figures and above tables. The numbering system is independent for figures and tables (the number increases with each figure, starting at 1 from the first figure shown; the number increases with each table, starting at 1 from the first table).

If plagiarism is intellectual theft and counterfeiting, falsifying data is scientific fraud and can kill a career in university or industry. Engaging in either practice—even as a student—is a very serious offence. If your data don't seem right, take a deep breath and decide whether you should try again to collect data (provided you have the time and resources) or if you just have to present what you have. Honesty requires that you include the first set of data as well. In your discussion you can mention any strange attributes of your data set, but try not to turn it into a rant about how you've failed! It's much better to do the best you can with what you have than to risk academic misconduct and damage your reputation.

Jan Pechenik put it nicely in his 2013 book, *A Short Guide to Writing about Biology:* "Decide exactly what you want your reader to see when looking at each graph or table, and then stick the reader's nose right in it." You need to put text in the Results section that helps guide the reader. You've had more time than he or she has to identify the trends, so share your expertise. Be clear about what the data show, but resist the temptation to explain what they mean. That's for the next section!

 Go to **science3.nelson.com** for a discussion of line graphs and vertical bar graphs, and an explanation regarding the kinds of data that each is best suited for. There's also a video tutorial on how to use Microsoft Excel to create these graphs (for both Macintosh and Windows computers).

DISCUSSION. The careful organization of your experimental report means that this is the section your readers will turn to in order to see how you analyze and interpret your data. The results section has the processed data (your averages and graphs) and a display of the trends. However, in the discussion section you'll relate your data back to your proposal in the introduction. If you made a hypothesis, did your data support it? You need not stick to past tense in this part. If there are future implications, then it is tricky to use the past tense while still sounding intelligent! Try to present the material with an active voice to make it more interesting, but keep it professional.

Note that in science, you can never be 100 percent certain. Your experiment cannot prove that your hypothesis is correct, but it can support it. Science works by continually testing what we know experimentally. Many journals only cautiously accept experiments that contain negative results because they don't necessarily disprove a hypothesis, so there's a bias in laboratories to try to publish work that demonstrates connections between the independent and dependent variables. Students also view experiments that don't show "what they are supposed to" as a demonstration of failure. True, if you're given all the details and follow instructions properly, the results should be as expected provided the laws of nature haven't changed, but "cooking your data" to show the proper effect is considered serious academic misconduct. If you're creating your own experiment and protocol, the chances of your data not being as you expect are even higher. Report your findings honestly anyway. Scientists need to be careful of their integrity, and honesty in reporting is critical. Besides, careful innovation and data collection can turn aside "common knowledge" and be one of the most exciting and compelling reasons to participate in the scientific enterprise!

Some high school science programs tell students to list their sources of error. Although this is a good practice in general, and it belongs in the discussion, it's important not to create a list of "all the terrible things you did experimentally and why the results aren't to be trusted." You should of course minimize sources of error when doing the experiment, but any assumptions you made regarding the apparatus and samples should be revealed. Use this as an opportunity to suggest refinements in the protocol for you, or perhaps the reader, to try out.

Don't reiterate all the ideas from your introduction, but do reflect back on what you said there. Clearly identify whether your hypothesis was supported or contradicted. Feel free to speculate on what the experiment might mean to the discipline. Is there any potential application of what you found out? Your instructors will be impressed if you show some insight here and relate the information back to the discipline in appropriate ways.

If you're a scientist by profession, you'd use this section to show how you could use more funding to make your own work even stronger. Good writing is critical to career advancement. The discussion is how you convince your reader that you know what you're talking about, how you contributed to your field, and why you should be supported to do further research.

And don't forget referencing! Although this section is where you do your analysis and put your own ideas forward, you are still obligated to relate this work to that of others. If your work confirms the findings of other labs, name them. If it contradicts their work, make it clear. If your ideas are working at the fuzzy edges of what is known, explain how your work fits into the theories and expectations of other people working in the field.

 DANGER ZONE

When actually "doing science"—exploring whether a premise is plausible through experimentation—students often worry that the experiment "didn't work" when they see that their hypothesis was not supported. Science isn't about proving you're right. If a test of your logic isn't supported by the results, you've still made a contribution. In such cases you need to look at how the experiment was done to see if the methodology was poor or whether some unforeseen factor altered the data. The popular TV show *MythBusters* assesses claims that people make about urban legends or anecdotal stories. They design their experiments after reasoning through the physics and chemistry of the phenomenon they're investigating. After running the test for the first time, they decide whether the data are reliable or not, then refine their protocol. At the end they decide if the legend is "busted." Their so-called negative results are useful because they identify bad reasoning or unlikely rumours.

REFERENCES/LITERATURE CITED. The purpose of this section should be obvious. We've covered plagiarism, and this is where some students get into difficulty. When you use the work of others, you need to help your reader find the source so they can receive the same inspiration as you. Also, to be frank, scientists need to be transparent about where they get their ideas, and it's appropriate to allow others to check up on you to see that you're using the work of others properly and in context. Science is self-correcting, and when practitioners are found to be dishonest in how they present their work, they are identified and their work is *retracted* (not used as "official information"). Your introduction, method, and discussion should contain citations that link up to the references you list in this section.

You'll find that this section is the one that requires the most attention to ensure it's in compliance with the writing style you're asked to use. Luckily, there are many resources online that help you use the proper style, and librarians are a great source of help for this.

Referencing may be done as author-year in your text, or you may use in-sequence citation (refer back to Figure 8.1). If author-year is used, then the references are alphabetized by the primary author's last name. Author-year is helpful when you want to look up a particular author. It can be a little tricky because the final name in a list of authors is often the primary investigator, the person who is supervising the work. Therefore, the work that comes out of a particular lab may be distributed throughout the list. In-sequence citation means that you need to keep track of when each reference is used, but tends to group the work done by a particular lab, or by labs all studying the same things. The drawback of this style is that if you rearrange the text of your introduction, method, or discussion, you need to rearrange the order of your references here.

Referencing Styles. You might find it useful to explore a resource such as EndNote or RefWorks. There's even a referencing tool for Microsoft Word (Write-N-Cite) that might be helpful for making a bibliography of works cited. Some universities have purchased site licences for these products, and your librarians can show you how to use them. Using these takes another skill that will require time to develop, but you will also save time if you have to switch styles. If your career involves much scientific writing, you'll do well to be able to change your style when required: It's common practice for a researcher to submit work to a high-prestige journal with a high rejection rate. If a paper is declined, it will be reformatted for another journal with the goal to have it accepted there. Sometimes an experiment will be reformatted several times for several journals before it is accepted.

Be careful of a common trap. Sometimes a student will pad the paper with a bunch of references collected based on their titles. You must actually *read* the parts of the papers you cite. If the paper you want to reference acts only as a connection of a fact to your work (i.e., if it doesn't contain the original fact you want to support but only references it in an earlier paper), then you must get the original reference and use that. Science is supported by a web connecting the work of a myriad of researchers. Referencing a paper that references another that references a lab that had the original fact creates a weak link between nodes. There's too much room for errors of interpretation, and it's bad practice to use weak or peripheral references.

APPENDICES. This is not a required section and will hold information that your instructor requires or data that don't fit properly into the other sections of the manuscript. Raw data are not normally required by journals, but your laboratory instructor may want it. Don't include this section unless you think it's really required. Check with your instructor to be sure she or he doesn't interpret your inclusion of extra information as simply "look at all the work I've done!" As an instructor who has marked many manuscripts, I can say that excessively long submissions with pointless extra attachments sometimes make me wonder if the student understood the task. If the information is important, it will almost always fit into the regular categories.

❓ Did you know?

Isaac Asimov was a famous essayist and science fiction writer. When he was about to write his dissertation for his PhD in biochemistry, he had been writing professionally for almost a decade. He worried that writing fiction might have damaged his ability to write academically. He did a spoof paper on the chemical nature of a fictitious compound called "thiotimoline" for practice. It was to be published under a pseudonym in case his examiners were to take offence if they thought he was trying to mock the stuffy writing of researchers of the time.

The Order of Writing

It is natural to think that the best way to write a piece of work is to start at the beginning and go to the end. The majority of science students probably begin with a title, make an abstract, then an introduction, and so on, until the paper emerges. A blank white screen or piece of paper can be intimidating. Many people find writing to be a process that begins slowly, then picks up momentum. For this reason, you might try a trick that can help you get started on writing a paper for science (see Box 8.4).

BOX 8.4   A Trick: Don't Write in Order!

Because you've done the work already and it's fresh in your mind, write the Method section FIRST, then proceed through the other things you've done. Here's a suggestion for the order in which you should do the actual writing (just remember to put them in the proper order before submitting!).

- Method

- Results

- Discussion and Introduction (You decide which is easier to start with, but you will have to tinker with these alternately to make sure they form two parts of the same thing! For example, you can refine your hypothesis a bit to accommodate changes you made in your experiment and make the implications of your results more clear.)

- References

- Abstract (Referring to the already-written manuscript for this summary makes it easier!)

- Title (Strangely enough, this is easiest at the end! You'll have all the important ideas sorted out and can make a descriptive, complete title with little effort!)

THE OUTLINE. An excellent writing aid is an outline. Make up a list of the categories presented here and jot down notes as you go along. That way, you won't forget the parts. Don't worry about complete sentences! The outline tool in Microsoft Word is great for this: You can group and reorganize your thoughts without rewriting everything. A lot of people view outlines as taking up too much time. The fact is, the effort you put into an outline will save you time later on. A good, careful outline *will* take time, but it's time you'd otherwise have spent writing down potentially disorganized sentences for your manuscript.

Be sure to point out the facts that underlie your study in the introduction part of the outline. You should do your research before you do the experiment, because, after all, you need to understand the system you're working in and be able to support the reasoning behind your hypothesis. Serious and competent scientists never do experiments on the fly. List the references here, too, for inclusion in your final work.

You should also think ahead to how you'll present your results. Will you use a bar graph? A line graph? A table? You'll want something that shows how your treatments affected the data. It's a good idea to plan this before you do the experiment and write it down. You can always change your mind about the presentation style later.

THE EXPERIMENT. There's not much to say here. With your preliminary research done and organized via your outline, you should have a good idea what your measurements will show. Science is rarely as simple as "let's see what happens when I do this." Properly done science is a deliberate assessment of a system. A well-reasoned protocol, with similarly well-reasoned controls, test material, and measuring devices, should result in data that—if unexpected—can be tested further to push back the boundaries of our ignorance of the physical world.

Don't forget to plan to do replicates so that you can see if the data you obtain are consistent! In addition, replicates indicate how reproducible your findings are.

THE MANUSCRIPT. Don't start with the title. Instead, get going on the method! After all, you've already done this part: You just have to explain what you did. Once you've built up steam, go to the results. They naturally flow from the method. Consult your outline to make sure you're not forgetting anything, but feel free to deviate from the outline if you think your work is better for it. The introduction can come near the end of your writing. This will give you the opportunity to ensure that you set the stage to allow your audience to understand how you did the work and why you chose the measurements you made (both of which are in the method section). Your hypothesis will be at the end of your introduction, and you can verify that your results section addresses that hypothesis.

Once your rough draft is done, think about how you can extend the work. Science is driven by inquiry. What new experiments could sort out conflicting data or make your data more clear? Are there alternative approaches you can take in your tests that would confirm that your results are real? It's a great idea to show your engagement with the material by suggesting new approaches or ways to get around limitations you recognize in the approach you used this time.

Once you've got the major sections done, consider doing the abstract. If everything else holds together well, you can now create a summary—your abstract. Consider writing one sentence on why you did the experiment, another sentence on how you did it, and a final sentence that points out one or two important findings from your work. There is no set number of sentences for your abstract, but you need to keep it as brief as possible, while still remaining informative for potential readers to decide if your work is worth reviewing in detail.

Once you get started, you'll find that later sections are easier to write. At the end, consider a meaningful title, then check that all your references are found in your text and that all the citations in your text are backed up with references. Double-check the references for accuracy! It's very frustrating not to be able to find a reference that intrigues you after reading a paper. In fact, misrepresenting references (intentionally or not) is considered by many to be plagiarism.

Then revise! Don't ever hand in your first draft. You should budget your time so that you can give yourself a few hours off before **proofreading** your work. If you find yourself changing parts of your work, don't feel like you did a bad job! Students who excel in science tend to take a few drafts to get to their best, most complete work.

8.6 OTHER FORMS OF COMMUNICATION

You may ask, "What if I'm NOT writing an experimental report?" Here are some tips on other forms of communication.

Research Paper or Essay

You might be doing a research paper instead of a formal laboratory report. Good news! This kind of paper only has four parts—Title Page, Abstract, Main Body, and References— instead of the greater number of sections for an experimental report. You're unlikely to have a hypothesis and maybe not methods, but you need to be clear and coherent, just as you are for a lab report. Refer to the sections above for Title Page, Abstract, and References.

The main body will have an introductory paragraph—or even a separate introduction—that fulfills the role of the introduction of the lab report. You'll orient your reader to the issue at hand and give a glimpse of the arguments you'll make later on. Divide the main body into sections to help your reader. These sections—level headings—are described on page 288 under the heading "Sections of the Manuscript."

Referencing for the research paper will be just like that for the laboratory report. Be sure to link the important points in the main body with the individual references you listed in the references section, and be sure to adhere to the specific referencing style suggested by your professor.

Oral Paper

You might have the opportunity to speak to a group about your work. This is often called "presenting a paper" and is usually considered a prestigious opportunity. Many conferences select their oral papers through a competition. Prospective presenters will submit a copy of their title and abstract to the conference committee, who then review all the entries and decide which ones to accept and which to decline. Offers to resubmit declined papers as posters are sometimes provided.

Just like any work you do, you'll want to be clear and confident in the way you present. You might wish to use Microsoft PowerPoint or some other presentation software to help you with your talk. Remember, though, YOU are the star. Don't let your presentation overwhelm the audience. Speak slowly and clearly, and by all means have notes available to yourself—don't read the slides out to the audience (they're very intelligent and can do that for themselves). If you rehearse enough, you won't need your notes, but don't feel bad if you end up using them. If you have notes, try to look up at your audience to make sure they're following you. This also creates a connection between you and your audience. You can also make it clear at the start of your talk whether you would like questions as they arise or if you want them at the end of your talk.

If you use PowerPoint, use as few words as possible. Too much text makes it difficult for your audience to read, particularly in a large venue. Also, avoid themes and transitions if possible. You don't want to be boring, but a lot of the time people dress up bad work with glitter, or in some cases overwhelm good work with busy slide transitions. Don't get your slide tips from Microsoft or a site that helps business people get sales. You'll get tips about how to dress things up or how to hook your audience early. Instead, your goal is to illuminate—be clear and let your data get your audience excited. Use large font—around 44 points for titles and no less than 24 points for text. If you can, limit each slide to three points. They need not be full sentences. Use a lot of contrast between your text and the background, and also within figures. Some people advocate for a white background to avoid eyestrain for the audience. In addition, a white screen illuminates the room more to make it easier for people to take notes. Make it as easy as possible for your audience to gather the information visually while still being able to listen to your words.

Moreover, let your presentation come from you. The graphics support the speaker and not the other way around. Your audience members probably can't split their attention effectively, so simple, tidy graphics and a short, to-the-point delivery are most effective. Keep in mind that yours might be one of many talks that day, so you must keep to the point and drill it home. Don't present everything you've ever discovered. Choose something interesting and keep on topic.

As for the layout, you might choose to present it with the sections outlined in the APA experimental report format if you did an experiment and are talking about it. If you're presenting information you gathered during research, you might do it with an introduction, body, and conclusion. In all cases you would not bother with the abstract: It will probably be published in the conference resources.

Leave time for questions at the end of your talk! If someone asks you a question, repeat it loudly and clearly before answering. Lecture rooms are often designed with good acoustics only from the front of the room, so the audience member's question is likely to be too quiet for everyone else to hear. You can also thank your supervisor and assistants, and acknowledge your sources of funding, if any.

Poster Presentation

These are presentations made in a large area (often a hallway or an auditorium) where the work is displayed in large text on a poster (hence the name). The sizes of the posters can vary a bit, and in the past—and even today—a poster can be made up of a series of panels that are affixed to a wall, a cardboard backing, or in some cases a formal display. A common size is four feet wide by four feet high, but this may vary depending on the venue. The actual sizes are published in the "Call for Presentations" associated with the conference. For a school project, your instructor will tell you the required dimensions.

The sections will likely be organized much like either the APA essay or APA experimental report formats described earlier in this chapter. It's up to you how to do it. Be sure, like you would for an oral presentation, to use very large font: 44 points for headings and 24 points for general text is fine. It should be readable from a short distance away.

Don't be afraid to add images or attention-grabbing devices (sometimes a plant, piece of apparatus, or cartoon related to the study is helpful). Posters are all displayed simultaneously, and the audience will be wandering among the displays. Unlike an audience at an oral presentation, which elected to sit in the hall to see the talk, the poster presentation participants will only stop at posters that interest them. Usually there is some time set aside for the poster authors to wait by their displays to answer questions. This usually occurs after some period of previous display so eager people in the audience can have a chance to look them over in advance and be able to ask better questions. Colour is fine, as long as the information is easy to read. Just like the oral presentation, you might have all the sections of an APA experimental report or an essay if that's more appropriate.

These tips are very short because this book is intended to be a quick reference guide. You can get a lot more detail and advice in several writing guides. *A Short Guide to Writing about Biology*, by Jan Pechenik (2013), is very good and covers everything from job application cover letters to doing critiques. CSE style is covered in handbook form by *The Pocket Handbook for Biology* (Kirszner & Mandell, 2004), along with a large section on grammar and spelling. There are many other discipline-specific texts that you might refer to for more detail.

8.7 REVIEW OF YOUR WORK

This section covers two topics: proofreading and peer review. Proofreading is used to catch errors before you submit work. Peer review is the act of ensuring integrity of data and ideas before making them available to the general population.

Proofreading

You are probably your own worst proofreader. The act of reading isn't what it seems: Your brain processes the words automatically and tends to fill in the blanks for words or letters it's expecting. Before handing in written work, it's a good idea to find someone you trust (whose opinion you respect and whose integrity is such that you know they won't "borrow" your ideas and words for their own assignment). Have this person act as your reviewer. Ask her or him to look over your work carefully. A lab partner is particularly good for this because she or he was there and should know how the experiment was run. Be sure you and your reviewer speak briefly about plagiarism and agree that your work must be substantively different from your reviewer's when you hand it in. Also keep in mind a "looks good" or "nice job" comment doesn't change the paper for the better. Ask for honest, valuable opinions. Even details regarding what you did well are more helpful than an empty accolade.

During a busy term it can be quite tricky to get a fellow student to proofread your work. A friend or relative—even someone who's not taking the same course as you—is better than you trying to do this for yourself. A very cool trick if you use Microsoft Word is to break your text into a separate paragraph for every sentence and sort them alphabetically, and then read to make sure each and every sentence makes sense. By removing the context, your brain is less inclined to "fill in" words that aren't there. Make sure each sentence is well constructed. Here is a bullet list for these steps using a Microsoft Windows computer and Microsoft Word 2007. Or you can watch a video at the *Science*[3] website to show how to do the procedure:

- Be sure to save a fresh copy of your document! You will be creating a new one that will destroy a lot of your carefully planned organization.

- Go to the very beginning of your document and choose "Replace" from the "Editing" option.

- For "Find what:" enter "." (a period … no quotation marks)

- For "Replace with:" enter ".^p" (a period, shift-6, lower-case p ... again no quotation marks).

- Choose "Replace All." Close the Replace window.

- "Select all" by holding the control button down and typing "a." All your text will be selected.

- In the Paragraph box, choose "Sort" (it's a graphic with an A/Z ↓ in it).

- Read each sentence as an individual unit. Each should be self-contained and make sense on its own.

- Find the bad sentences in your original work and fix them.

- A low-tech way to do a similar action is to read each sentence starting from the END of your document. Reading backwards takes the sentences out of context so you can focus on whether each sentence is clear and complete.

It can't be stated enough that your work should *never* be published (or submitted to your instructor) as a first draft. Realistically, you should go through four revisions—or more— to make your writing clear, concise, and incisive. Leaving your work to the last minute is a good way to underachieve. Keep in mind that the same time and class stresses you have are also experienced by your colleagues. You'll be doing yourself a favour by getting at your work early—long before the deadline—and giving yourself at least a few hours of "cooling off" before you re-read your work. If you give yourself a decent interval, you'll be amazed at the creative ideas for layout and presentation you'll come up with when you revisit your work!

Peer Review

ITS ROLE IN SCIENCE. Peer review in the sciences adds credibility to what has been published. A paper must be open and transparent, and it needs to include good references that support it within the discipline. When you publish—particularly in a journal—you need to allow others to test what you've done and see if it's a phenomenon that is repeatable using their own equipment.

HOW IT IMPROVES YOUR WORK. You must normally lay out all your logic and all your steps so others can ensure that the data make sense. Even as a student, you'll get good practice if you write your work as if it's intended to be published. Good practices don't just happen through sheer will ... they take ... *practice*.

8.8 CONCLUSION

Communicating science is a critical part of a scientist's job. Without communication, science is powerless. Communicating effectively takes a lot of effort, and the first few attempts are generally imperfect. In fact, seasoned investigators still find improvement the more they engage in writing and other forms of communication. Some educators, students, and researchers find that keeping a blog allows for excellent opportunities to improve. Seek out honest criticism and be sure to get started on your work early enough to allow time to incorporate suggestions. If you become proficient at writing, you are exceedingly more likely to be offered opportunities for scholarship and in your career! Good luck!

FURTHER READING

Angeli, E., Wagner, J., Lawrick, E., Moore, K., Anderson, M., Soderlund, L., & Brizee, A. (2010). *Purdue Online Writing Lab.* Retrieved January 4, 2012, from http://owl.english.purdue.edu/owl/resource/560/01/.

Chapman, O., & Fiore, M. (2001). *The White Paper: A Description of CPR.* http://cpr.molsci.ucla.edu/cpr/resources/documents/misc/CPR_White_Paper.pdf.

Gillis, C. (2011, November). When professors plagiarize. *Maclean's, 124*(43), 84.

Gladwell, M. (2008). *Outliers: The Story of Success.* New York, NY: Little, Brown, and Company.

Kirszner, L., & Mandell, S. (2004). *The Pocket Handbook for Biology* (2nd ed.). Boston, MA: Thompson.

Leamnson, R. (2002). *Learning (Your First Job).* http://www.udel.edu/CIS/106/iaydin/07F/misc/firstJob.pdf

Miah, M. D. T., Khan, M. A. I., Ahasan, H. A. M. N., Das Gupta, R., & Alam, M. D. B. (2011). Plagiarism in medical science—The forbidden fruit. *Journal of Medicine, 12,* 160–162.

Pechenik, J. A. (2013). *A Short Guide to Writing about Biology* (8th ed). Toronto: Pearson Education.

Teitel, E. (2011, November). Accused. *Maclean's, 124*(43), 82.

"U of Alberta dean accused of plagiarism apologizes." (2011). http://www.cbc.ca/news/canada/edmonton/story/2011/06/13/edmonton-dean-apology-plagiary.html.

Walker, J. (2010). Measuring plagiarism: Researching what students do, not what they say they do. *Studies in Higher Education, 35,* 41–59.

Job, Scholarship, and Post-Graduate Applications

> "*Being a graduate student is like becoming all of the Seven Dwarves. In the beginning you're Dopey and Bashful. In the middle, you are usually sick (Sneezy), tired (Sleepy), and irritable (Grumpy). But at the end, they call you Doc, and then you're Happy.*"
>
> RONALD T. AZUMA, 2003

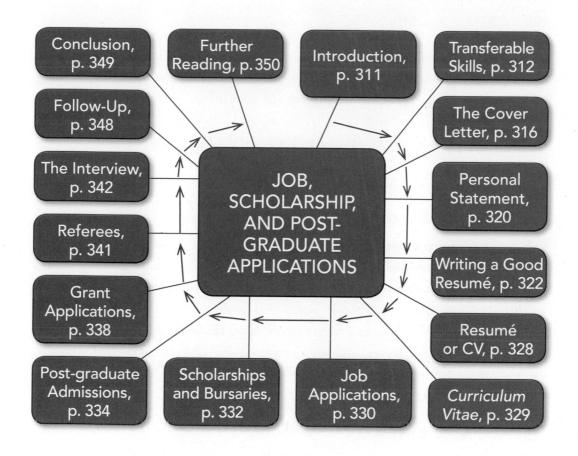

9.1 INTRODUCTION

The opening quote of this chapter refers to the Seven Dwarves of Snow White fame and is written about graduate students—but it applies equally well to undergraduates. Think back to the beginning of first year when you didn't know your way around the university and weren't sure of all the rules and regulations. Did you feel a bit Dopey or Bashful? Certainly Sneezy, Sleepy, and Grumpy describe many students in the middle of exams and final papers! I guess the only dwarf you can't really emulate until after graduate school is Doc!

This chapter is aimed at helping you prepare for the toughest thing you will ever have to do in or out of school—convince people that they want you and, even more than that,

that they want to give you money! Whether you are looking for a job or trying to get into graduate or professional school, a big part of what you have to do is self-promotion. If you want to think of this as a sales pitch, then that works too. If you are trying to get money from a funding agency for a research project or a scholarship, you have to make an even better sales pitch. You, and your project, have to stand out—in a positive way—from all the other students who are competing with you for the same position, scholarship, or funding grant.

Why Are You Reading This Chapter?

You may be a first-year student thinking about a summer job that doesn't involve fast food, or perhaps you are a fourth-year student about to graduate and you're looking at graduate school or a job in the private sector. Perhaps you need to write a grant application or a 750-word essay on a scholarship application. Whatever your ultimate goal, you need some advice and some clear-cut examples of how to make you the best contender, the one everyone will want to have as part of their company or program, the one everyone will think should be funded because you and your project are just that great.

General Principles versus Specific Requirements

Every school, program, company, or project will have specific requirements for its application process. This chapter can't cover all of these, so I'll focus on some general principles that you can apply to your specific circumstances.

9.2 TRANSFERABLE SKILLS

Before even beginning to discuss what you need to do to get a scholarship, a job, or a position in graduate or professional school, let's talk about what your skills are. No matter whether you are fairly new to school or almost finished, you are learning things in your classes that will help you succeed at whatever comes next. By this I mean that not only are you learning science but you are also developing abilities that will be useful to you as a human being, a lot of which can be used in your cover letters, resumés, personal statements, and all the other things that will be discussed in this chapter. In Box 9.1 is a list of several skills that can transfer from school to other areas of your life. This list is not complete but should get you thinking about what you can do that could be added to your other more obvious abilities.

BOX 9.1   Transferable Skills in Science

Scientific Skill	Place You Learned It	How the Skill Transfers
General science skills	Science courses—lecture, lab, and exams	Appreciate the complexity of the Universe and of life. Think critically to evaluate data and evidence. Have a toolbox full of methods for studying a new topic.
Scientific writing	Literature reviews Lab reports Lab notebook Mini-essays (1–5 pages) Major papers Poster presentations Tests/exams	Communicate in a variety of forms. Know appropriate citation and referencing methods. Give clear and accurate accounts of a topic in writing.
Oral communication	Solo presentations Group presentations Discussions Debates	Give a clear and accurate verbal report. Facilitate a discussion.
Numeracy	Collecting data Analyzing data Interpreting data Representing data in graphs Statistical analysis Mathematical modelling Evaluating conclusions derived from data	Understand sampling and sampling errors, calibration, and types of error. Obtain appropriate data. Interpret and present data. Perform statistical analysis of data. Use data in various forms.

Continued

Scientific Skill	Place You Learned It	How the Skill Transfers
Computer skills	Exposure to, and use of, a variety of software programs	Use computers and various types of software to solve problems and work with data. Use computers to do literature searches. Use computers and diverse types of software to communicate.
Experimental design	Developing hypotheses Performing hypothesis testing Designing methods Following methods Analyzing experimental outcomes	Design, plan, and carry out an experiment. Evaluate results and modify the experiment.
Lab skills	Employing general lab techniques Observing experiments Recording data	Be comfortable in a lab setting. Learn new techniques as needed. Keep a record of work done.
Field skills	Observing in the field Collecting specimens Experimenting Identifying organisms in natural habitats	Work in the field solo or as part of a team. Obtain and analyze data. Follow legal and ethical requirements.
Proficiency with scientific equipment	Using various instruments in labs	Work with common equipment in your field. Learn to use specialized equipment.
Literature search and information retrieval	Completing background research for lab reports Studying for exams	Read and evaluate literature critically. Discern between poor, good, better, and best sources of information.

Continued

Scientific Skill	Place You Learned It	How the Skill Transfers
Literature critique	Critical reading Critical reasoning Assessing the literature Evaluating the literature	Discern between poor, good, better, and best sources of information. Analyze and summarize information.
Problem solving	Completing problem sets Participating in group work Doing "real" labs (i.e., non-cookbook labs)	Apply knowledge to solve problems individually or as part of a team.
Decision making	Participating in lab sessions Participating in group projects	Work alone or with a group to set goals and make decisions.
Leadership	Leading group projects Leading group work in class	Take the lead in a group while respecting other peoples' opinions. Keep others on task. Work with people from other backgrounds.
Teamwork accountability	Participating in group projects	Work as part of a team and take responsibility for your part. Evaluate your work as part of group.
Organizational skills	Obtaining a degree Completing assignments on time Studying for exams in a timely fashion	Work independently. Manage your time effectively. Set goals and work towards achieving them.
Ethics	Taking an ethics class or discussing ethics as part of a class	Identify, communicate, and debate ethical issues in science.

Source: QAA, Subject benchmark statement—Biosciences 2007. Retrieved from http://www.qaa.ac.uk/ Publications/InformationAndGuidance/Pages/Subject-benchmark-statement-Biosciences.aspx, 2007; A. Jones, R. Reed, and J. Weyers (2012). *Practical Skills in Biology* (5th ed.). Toronto, ON: Pearson Education.

9.3 THE COVER LETTER

There are several things that you need to keep in mind when writing a cover letter. First, keep it short, but cover all the essentials. This means you will probably need to write it more than once and then rewrite it to cut out the least essential bits and pieces. Your letter needs to fit on one page—you can do this if you write it carefully. Remember that the people reading these cover letters may be going through dozens—if not hundreds—of applications, so being concise is really important. The other crucial idea to keep in mind is that this letter is the first thing that will be seen by your prospective supervisor or funding agency; if it is not well written, nothing else will be read. (Refer back to Chapter 8 for a reminder of the importance of good communication.) You want to catch the reader's attention in a positive way, not by your grammatical and spelling errors! I have included a sample cover letter in Box 9.2, but you can also find some really great information on the Government of Canada website at www.youth.gc.ca/eng/topics/jobs/cover.shtml.

BOX 9.2   Sample Cover Letter

15 Waterloo Road
Metropolitan, Saskatchewan
S0S 0S0
4 July 2012

Mr. Walter Langorski
Human Resources Manager
Astrobe Laboratories Ltd.
1632 Bayonne Street
Two Planes, Ontario
M0M 0M0

Dear Mr. Langorski:

I am writing in application for the position of Junior Technician with Astrobe Laboratories Ltd. I think the water-quality tests you perform for industry are very innovative, and I would love to have the opportunity to be associated with such a forward-looking company. I have enclosed my resumé for your consideration.

Continued

Through my studies in the Environmental Chemistry program at South Atlantic University, I have developed a number of lab skills that I believe will be useful to your company. These include Soxhlet extractions, the usual distillation and titration techniques, and lots of experience in structure determination of unknown compounds. I have also had the opportunity to work with instrumentation, including both hydrogen and carbon NMR, ICP-MS, GC-MS, Raman, FTIR, and UV-visible spectroscopy. Writing weekly laboratory reports has helped me improve my communication skills.

As the team leader for a successful undergraduate research project, I developed organizational skills while ensuring everyone in the group had their work done by the required deadlines and preparing our poster for the Student Research Day at South Atlantic University. I am really good at both following instructions and being a self-motivated worker, as the circumstances require.

I would like to apply my laboratory and leadership skills to this position with Astrobe Laboratories Ltd. I hope you will give me an opportunity to meet you in an interview. Thank you for reviewing my application.

Sincerely,

James Hurdsman

James Hurdsman

Not only is a well-written cover letter important so that you can get the rest of your application seen, but most employers see good writing as an indication of good thinking. Before you begin writing, ask yourself: Why do I want this position? What skills would be most useful for this position? Which of these skills do I have? How can I highlight these skills and my other strengths in my cover letter? What else should my cover letter contain? See Box 9.3 for a list of the basics that must be in your cover letter.

BOX 9.3 Cover Letter Checklist

- Clearly state the position that you are applying for.
- Show how you are qualified for the position (i.e., draw the reader's attention to the aspects of your resumé that make you the person they want).
- Show that you understand what the position entails.

Continued

THE COVER LETTER

- Show that you are knowledgeable about the company/program/department—go on their website and do some research before you start writing.
- Show how you will fit into their departmental or corporate culture.
- Show how you will be a benefit to the company, department, or program.
- Show that you can communicate clearly and well.
- Show that you are mature and capable, but don't come across as arrogant!

Before going on, I have one comment to make about the last point in Box 9.3. It is inevitable that different application reviewers will have different preferences. That is, some will like the confident approach, while others may dismiss it. It is challenging to find the right balance, but usually a quiet confidence in your abilities and some examples of why you feel confident will hit the right note. This applies to both the letter and the interview.

Once you have written your letter, have someone proofread it for you. Career counsellors, parents, and/or instructors can usually be conscripted for this job. Be aware of your spelling and grammar and use the spell-checking feature that is standard with your word-processing software. Do keep in mind, as always, that a spell-checker cannot think for you; all it checks is spelling. You need to make sure, especially if a word can be spelled in more than one way, that you have the correct word for the context. (Box 9.4 has a poem written in 1992 where every word is spelled correctly but many are not the correct words for the context.) Also keep in mind that appearance is very important; keep it neat, professional, and in a formal business format. Print your letter on decent-quality paper but don't go overboard—if the paper is too fussy or expensive, that can trigger some negative responses as well—and never use coloured or scented (yes—some people actually do that) paper.

BOX 9.4   Candidate for a Pullet Surprise by Jerrold H. Zar

I have a spelling checker.
It came with my PC.
It plane lee marks four my revue
Miss steaks aye can knot sea.

Continued

Eye ran this poem threw it,
Your sure reel glad two no.
Its vary polished inn it's weigh.
My checker tolled me sew.

A checker is a bless sing,
It freeze yew lodes of thyme.
It helps me right awl stiles two reed,
And aides me when aye rime.

Each frays come posed up on my screen
Eye trussed too bee a joule.
The checker pours o'er every word
To cheque sum spelling rule.

Bee fore a veiling checkers
Hour spelling mite decline,
And if we're lacks oar have a laps,
We wood bee maid too wine.

Butt now bee cause my spelling
Is checked with such grate flare,
Their are know faults with in my cite,
Of nun eye am a wear.

Now spelling does knot phase me,
It does knot bring a tier.
My pay purrs awl due glad den
With wrapped words fare as hear.

To rite with care is quite a feet
Of witch won should bee proud,
And wee mussed dew the best wee can,
Sew flaws are knot aloud.

Sow ewe can sea why aye dew prays
Such soft wear four pea seas,
And why eye brake in two averse
Buy righting want too pleas.

Source: J. Zar (1994). "Candidate for a Pullet Surprise," *Journal of Irreproducible Results, the science humor magazine, 39*(1), 13. Retrieved from http://www.jir.com.pullet.html

A Few Last Suggestions That Can Really Make a Difference

a) Tailor each cover letter to the position for which it is being prepared. Don't use the same letter over and over—you need to be specific!

b) Address the letter to the appropriate person—if that is not given in the job posting, do some research. Make a phone call if necessary! Find out if the correct salutation is Dr. or Mr./Ms. (make sure you know whether the person is male or female). Be sure you spell the name correctly.

c) Don't make general, unsupported statements. Give some details that show what you are capable of.

d) Do not simply repeat the job ad word-for-word in the cover letter.

e) Avoid using the passive voice. This is hard because many of us have been taught to write this way for lab reports and scientific papers. Try using "I did," "I led," "I created."

f) Sign your letter with your name, not a nickname.

g) Your email address should look professional. "sexy85@hotmail.com" can suggest to your prospective employer that you might pose a risk outside of your working hours.

9.4 PERSONAL STATEMENT

Personal statements are essays written in the first person that replace a cover letter if the company/institution you are applying to has an online application system. It may also be part of the application process for school or for a scholarship. Typically the website or application form contains instructions on the type of information the company/institution wants to see in the personal statement. Take the time to make a list of points that must be covered, write a draft, and have someone proofread it before you actually start typing on the website or filling out the form. Most of the comments made above regarding cover letters apply here. Box 9.5 contains a sample personal statement.

Keep in mind that just because the personal statement is sometimes submitted online doesn't mean you should be casual in your language or approach. As with a cover letter, this is your chance to make a sale: to impress your potential employer/supervisor. Try to be specific

I am applying for the position of Head of the Laboratory Safety Department because I believe I would be a good asset to the department and a strong leader. I have enjoyed the challenges as Deputy Head this year and believe that my strong administrative background helps me be effective in the job. I have previously served as Biology Safety Coordinator and worked closely with the previous department head, so I was familiar with many aspects of the job when I stepped into it. I have enjoyed the increased opportunities to lead the department in a very exciting time as we deliver, for the first time, a new program of laboratory safety at this institution.

The key strengths that I possess for success in this position are a great liking for the people I work with, deep integrity, and a strong work ethic. I think I have shown over many years at this institution that I am an innovative thinker and good at motivating others to move forward, even when confronted by unknown territory. A project I have greatly enjoyed working on is designing the new Hazardous Wastes Disposal Unit to serve the needs of this institution well into the future. I was instrumental in developing a new organizational system for assessing risk and obtaining funding from the government of Hidalgo for new instrumentation to track hazardous vapours to their source, and have encouraged my colleagues to develop waste-disposal tracking sheets. In addition, I have involved our colleagues throughout the institution by pioneering the idea of waste-disposal systems that could be used by researchers from various disciplines. I am excited by the new Safety Assessment Seminar that we have developed and are piloting for the first time this month.

I always try to lead by example, never requiring someone to do something that I would not or am not already doing myself. I believe that our department often leads the way in embracing change, and I feel strongly that as head I would con-tinue to encourage this in a well-thought-out and mutually-agreed-upon fashion. I believe arriving at consensus in departmental decisions is important and plan to continue to build on the foundation laid by our previous head in this area. I know that I will sometimes make, and indeed have made, mistakes, but I am willing to acknowledge the mistakes I make and learn from them.

Continued

I trust that my dedication to the Laboratory Safety Department and this institution is evident to you and to the committee. I look forward to the opportunity to discuss my application with the selection committee.

about why you want to be at that institution or in that program. In other words, show that you have done some research on the place you are applying to and on how you would be a good fit. You are trying to let them know why you would be an asset and why they should consider your application above all the others they have received. Also, and most important, make sure your statement fulfills the directions given on what you are to discuss in your statement and how long it should be.

When you submit something online, keep a copy of what you sent by copying and then pasting it into a separate document, printing the page, or some other method. That way you can remind yourself, if you get an interview, of exactly what you told them. Last, and most important, do not hit "send" or "submit" until you have re-read your submission at least once and preferably had someone else read it as well. None of us needs to have that sinking feeling that comes with knowing you made a mistake that can't be quietly fixed.

9.5 WRITING A GOOD RESUMÉ
Paper Resumé

As with cover letters and personal statements, a resumé is a chance to show what you have accomplished to this point in your life and why you would be an asset to the company/school/organization. There are many acceptable formats, and most word-processing programs contain one or more templates for resumés. Keep in mind that you may need to tailor these templates to fit your specific needs. In addition, try to choose a template that looks professional to you (as usual, it is always a good idea to get a second opinion). Box 9.6 contains a sample resumé.

BOX 9.6   Sample Resumé

Yulissa MARKISON   805-216-5034

3-81 Line Way, Whalleye, Ontario L1T OS9   ymarkison@jiffymail.com

OBJECTIVE: APPLIED SCIENCES LABORATORY TECHNOLOGIST

SUMMARY:

- Multidisciplinary knowledge of biochemistry, microbiology, and molecular biology obtained through both college and university education and work experience.

- Prepared and calibrated standard samples. Operated the instruments independently.

- Working knowledge of techniques for cell biology, protein purification.

- Experience in collecting, analyzing, and interpreting data. Experience keeping a formal laboratory notebook following SOP and USP guidelines.

- Effective interpersonal, organizational, and project management skills developed through professional experience.

- Working knowledge of various data acquisition software applications and Microsoft Office.

- Total time spent in both college and university laboratories: over 700 hours.

PROFESSIONAL SKILLS:

Chemistry:
Solution preparations, titrations, and gravimetric analysis. Product content analysis, as well as instrumentation techniques such as spectrophotometry (UV/Vis, IR, NMR), chromatography, flame emission, atomic absorption, HPLC, and GC.

Biochemistry:
Preparation of agarose gels and SDS–PAGE gels from scratch, protein purification by chromatographic techniques (size exclusion, ion exchange, affinity), DNA extraction and PCR, protein assay/characterization, SDS-PAGE, Western blotting, ELISA.

Microbiology:
Tissue culture aseptic techniques, cell passage, media preparation, dilutions, plate pouring, plate counts, spore transfers, plate streaking, gram staining. Environmental analysis as well as product analysis following USP guidelines.

Continued

Statistical Methods in Quality:
Knowledge of bioinformatics and modelling software, GLPs, GMP, WHMIS, IUPAC, BIOSAFETY, and other regulations. Authored reports and summaries, performed data entry.

Technical Communications:
Writing technical reports, standard operating procedures, certificates of analysis, formal laboratory notebooks for all disciplines of study.

WORK-RELATED EXPERIENCE:
Laboratory Technologist Assistant   April 20XX–September 20XX

Eastern University

- Data entry

- MSDS project management

- Inventory management

- Supply management

- Shipping and receiving

- Interpreted data

- Wrote technical reports

Quality Inspector   April 20XX–September 20XX

The Inspection Group

- Inspected manufactured parts to ensure quality standards are met.

- Managed all quality-related documentation.

- Oversaw staff schedules, balanced overtime, and submitted attendance records for payroll.

- Supervised a team, acted as a liaison between quality assurance staff and the project manager.

- Participated on internal auditing team to maintain the quality standards of an ISO-9000 certified company.

EDUCATION:
B.Sc. Cell and Molecular Biology   September 20XX–April 20XX

Eastern University

Continued

Biotechnology Technologist   September 20XX–April 20XX

Northern College

3-year Ontario College Advanced Diploma

Graduated with Honours

Graduated on Dean's List

References are available upon request.

Your resumé should include information about your education, relevant job experience, relevant research experience, goals, and personal interests. In each section of the resumé that involves dates, you should put information in chronological order from most recent to most distant, for example, 2012, 2011, 2010. If possible, keep it to one page and try to follow a logical order for the various sections of information.

When presenting information, use action verbs such as "led," "did," "created," and "formed" to describe accomplishments. You can emphasize important bits using boldface, italics, bullets, or some other reasonable method, but try to be consistent in your usage or it will look messy. As always, proofread for errors, have someone else proofread it, and print your resumé on good-quality paper.

Scannable Resumé

A scannable resumé is, as suggested by the title, designed to be scanned into a database. It should be set up in almost exactly the same way as a print resumé and should contain the same information. The two main differences between a scannable and a print resumé are that the scannable resumé should not be formatted (i.e., no columns, bullets, italics, or bold lettering), and it should contain key nouns or adjectives to describe accomplishments if they are not already in your regular resumé. These are in less and less use now due to newer methods for data transmission, but a sample is included here in case you need to make one (see Box 9.7).

6

BOX 9.7 Sample Scannable Resumé

Yulissa MARKISON

805-216-5034

3-81 Line Way

Whalleye, Ontario L1T OS9

ymarkison@jiffymail.com

OBJECTIVE: APPLIED SCIENCES LABORATORY TECHNOLOGIST

SUMMARY:

- Multidisciplinary knowledge of biochemistry, microbiology, and molecular biology obtained through both college and university education and work experience.

- Prepared and calibrated standard samples. Operated the instruments independently.

- Working knowledge of techniques for cell biology, protein purification.

- Experience in collecting, analyzing, and interpreting data. Experience keeping a formal laboratory notebook following SOP and USP guidelines.

- Effective interpersonal, organizational, and project management skills developed through professional experience.

- Working knowledge of various data acquisition software applications and Microsoft Office.

- Total time spent in both college and university laboratories: over 700 hours.

PROFESSIONAL SKILLS:

Chemistry:

Solution preparations, titrations, and gravimetric analysis. Product content analysis, as well as instrumentation techniques such as spectrophotometry (UV/Vis, IR, NMR), chromatography, flame emission, atomic absorption, HPLC, and GC.

Biochemistry:

Preparation of agarose gels and SDS–PAGE gels from scratch, protein purification by chromatographic techniques (size exclusion, ion exchange, affinity), DNA extraction and PCR, protein assay/characterization, SDS-PAGE, Western blotting, ELISA.

Continued

Microbiology:
Tissue culture aseptic techniques, cell passage, media preparation, dilutions, plate pouring, plate counts, spore transfers, plate streaking, gram staining. Environmental analysis as well as product analysis following USP guidelines.

Statistical Methods in Quality:
Knowledge of bioinformatics and modelling software, GLPs, GMP, WHMIS, IUPAC, BIOSAFETY, and other regulations. Authored reports and summaries, performed data entry.

Technical Communications:
Writing technical reports, standard operating procedures, certificates of analysis, formal laboratory notebooks for all disciplines of study.

WORK-RELATED EXPERIENCE:

Laboratory Technologist Assistant

April 20XX–September 20XX

Eastern University

- Data entry
- MSDS project management
- Inventory management
- Supply management
- Shipping and receiving
- Interpreted data
- Wrote technical reports

Quality Inspector

April 20XX–September 20XX

The Inspection Group

- Inspected manufactured parts to ensure quality standards are met.
- Managed all quality-related documentation.
- Oversaw staff schedules, balanced overtime, and submitted attendance records for payroll.

Continued

- Supervised a team, acted as a liaison between quality assurance staff and the project manager.
- Participated on internal auditing team to maintain the quality standards of an ISO-9000 certified company.

EDUCATION:

B.Sc. Cell and Molecular Biology

September 20XX–April 20XX

Eastern University

Biotechnology Technologist

September 20XX–April 20XX

Northern College

3-year Ontario College Advanced Diploma

Graduated with Honours

Graduated on Dean's List

References are available upon request.

9.6 RESUMÉ OR CV

Many people use the terms resumé and CV (short for ***curriculum vitae***) interchangeably, but they are not the same, at least not in North America, with the exception of Québec. Typically, a short and succinct resumé as discussed above is used for most job applications. A CV is much longer and contains relevant awards, conference presentations, fellowships, grants, licences, memberships, professional associations, publications, research experience, teaching experience, and so on. A CV is typically expected when applying for jobs in academia, research positions, or professional positions such as in law or medicine.

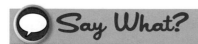

CURRICULUM VITAE: From the Latin words *curriculum*, meaning "course," and *vitae*, meaning "of life."

9.7 CURRICULUM VITAE

Your *curriculum vitae* (CV) will typically be much longer than your resumé, but no one is interested in fluff: Keep it real and on point. For example, don't say you taught a course when you guest-lectured once or twice. No one is ever impressed by padding. The following is a list of the sections that are usually found on a CV for an academic position in the sciences. Some of these headings are different depending on the exact purpose of the CV, but this is a general guideline that can get you started.

BOX 9.8   *Curriculum Vitae* Checklist

- Post-secondary education history
- Degrees
- Post-doctoral positions history
- Employment history
- Educational institution employment
- Other employment
- Awards and distinctions
- Teaching and other supervision activities
- Course development
- Coordination
- Student supervision
- Publications and other scholarly output
- Journal articles
- Books
- Professional communications
- Conferences
- Conference papers
- Professional activities
- Other professional activity

Continued

- Participation/membership activities
- Committee memberships
- Event participation
- Membership in scholarly societies or other organizations
- Administrative/organizational/executive activities
- Offices held
- Editorial activities
- Assessment/review activities
- Grant application assessment
- Other assessment activities
- Other (teaching, scholarship, service)

 TOOL KIT

A sample CV is not included here because CVs tend to be really long. The following website is set up by the Canadian government and contains an instructional webinar that could be helpful to you in preparing a CV: https://ccv-cvc.ca.

9.8 JOB APPLICATIONS

Getting a job involves a lot of hard work, and the approach you take depends to some extent on what type of job you are trying to get. For example, getting a summer job at your university is more likely to involve your knowing one of the professors well enough that he or she might mention the position to you. These type of jobs are rarely posted outside the university and in most cases are not posted at all but rely on word of mouth. If there is a professor with whom you would like to work, there's no harm in asking her or him if such opportunities are available. A respectful inquiry won't be offensive. However, this approach is not as useful when you are looking at career positions or a place in a graduate or professional program, because there are typically more formal application processes in place for these types of positions.

That being said, networking or making personal contacts with people in your field is always useful when looking for advancement. It is often through conversations with people at conferences and other meetings that you find out about available positions. You also have to keep in mind that by getting to know people and showing an interest in their work, you give them a chance to get to know you. For example, recently a fourth-year student I know well presented a poster at a conference and during the session was offered a position in a lab at another institution. Clearly, this doesn't happen every time someone presents a poster, but certainly taking advantage of undergraduate poster sessions and other such networking opportunities can be helpful in advancing your career.

However, keep in mind that networking might give you a foot in the door at best. Once you become aware of a position that interests you, you still need to approach your cover letter and resumé in a professional manner. You still have to show the prospective employer that you are worth his or her consideration.

The next most effective method of getting a job is searching the job ads, whether online, in the newspaper, or at job fairs. If you see an advertisement that interests you, do a little research before submitting your application so that you tailor it to the specific requirements of the job and personalize it by finding out the name of the person to whom your cover letter should be addressed. Obviously, the easiest (and probably least effective) thing to do is submit a cover letter with a resumé to companies that are doing something you are interested in but that are not currently advertising for a position. In most cases, this approach won't get you a job. Most companies identify the need for a new employee and investigate their budgets before advertising that they are hiring. One other suggestion: Before you actually start applying for jobs, call around to places that have the types of jobs you might be interested in, and ask if you can talk to someone about the qualities they would look for in a good employee. Promise you will only take 15 minutes of their time, and then keep your promise. At the 15-minute mark, say "thank you" for the insight they have given and stand up to leave. The person you are talking with will let you know if they are willing to spend more time with you or not.

9.9 SCHOLARSHIPS AND BURSARIES

Most institutions have a Student Awards Office (although it may be called something else at your institution). The best way to find it is to go to your school's website and search for **scholarships (entrance, external, national, and renewable)**. This will usually lead you to the information you are looking for, including eligibility requirements, application forms, and deadlines. There can be some confusion about the words *scholarship, bursary,* and *grant,* and there are some differences in how the words are used in Canada and in the United States. In the United States, Federal Education Grants are provided to individual students according to financial need, with the goal of helping students get to university who might not otherwise have the opportunity. In Canada, **bursaries** or grants for financial need are often given through the Student Loan system. That is, you apply for a student loan, which you are expected to pay back after you graduate, but some portion of it is remitted (doesn't have to be paid back) based on your financial need.

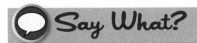

SCHOLARSHIP: A money award that recognizes academic excellence. Sometimes financial need may be a secondary criterion.

ENTRANCE SCHOLARSHIPS: Monetary awards you get when you first enrol at a school.

EXTERNAL SCHOLARSHIPS: Monetary awards available from organizations such as where your parents work or the rugby club you belong to. Not all schools will be approved for the awards, so double-check to make sure your school is approved by the organization offering the scholarship.

NATIONAL SCHOLARSHIPS: Monetary awards given to students all across the country.

RENEWABLE SCHOLARSHIPS: Monetary awards that can be renewed every year, often dependent upon certain criteria, such as maintaining high grades.

BURSARY: A money award based on financial need. Usually the application form includes a financial disclosure form. Bursaries may also require that students meet geographical or demographic requirements.

TOOL KIT

For students in Canada	www.scholarshipscanada.com/
For students in the United States	www.educationgrant.com/

Not all scholarships require an essay, but many do. The usual rules apply: Write your thoughts out separately from the online or paper form, re-read them after you are finished, proofread them, have someone else proofread them, and then write them out properly. Take your time and be thoughtful in your comments.

If you receive an award, make sure to send a thank-you letter to the donor. If there is an awards ceremony and the donor is present, be sure to give your sincere thanks to the donor. You may be required to give a speech upon receipt of the award. If so, be sure you know how long you are expected to speak for, write out your speech, and then have someone time you giving the speech. Cut out unnecessary bits if it is too long. If your speech is too short, think a little harder about how grateful you are for the award and how it will help you. As always, have someone else proofread your speech so that you don't embarrass yourself by using poor grammar. See Box 9.9 for a sample thank-you letter.

BOX 9.9   Sample Thank-You Letter

Mr. John Smith
111 Anywhere St.
Lane's End, Alberta, T0S 0K0
Canada
29 August 2012

Rebecca McGee
654 Mount Albert Road
Knoxville, Alberta, T1V 9V9
Canada

Continued

Dear Mr. Smith:

Thank you so much for the Smith Chemistry Award, which I received this year. It was such an honour, and I really appreciate the effort that your family takes to ensure that students each year will be aided financially in their studies.

This award will help me to purchase textbooks this year without having to take on a second part-time job. It was also really encouraging to know that someone felt that I have the ability to succeed in chemistry. Sometimes in the middle of the semester it is easy to lose sight of my goals, and this award supplied a great emotional lift as well as the more obvious financial help.

Since I really enjoy chemistry and hope to go on to graduate school once I have finished at South Atlantic University, the Smith Chemistry Award has provided added incentive to work toward my goals. As a first-generation Canadian, I have had to climb a lot of barriers to enter university, and I am very grateful for this award. Thank you again for the Smith Chemistry Award.

Sincerely,
Rebecca McGee
Rebecca McGee

A few hints about public speaking might be useful at this point. First of all, be sincere in what you say, but don't say anything that will get you emotional. Crying makes it very hard to give a speech! Second, watch the timing. You want to thank your donor—not bore them. Third, remember your audience is prepared to be pleased. They are there to congratulate you on your achievement, so take a deep breath, say what you have to say, thank everybody one more time, and then relax.

9.10 POST-GRADUATE ADMISSIONS

When you are applying to graduate or professional schools, the process will differ slightly for each school and for each type of program, so these are just general guidelines. You will need to get the specific instructions from the relevant website or admissions department. Graduate schools offer degrees that require a bachelor's degree as a prerequisite. The degrees offered are usually Ph.D., M.A., or M.Sc. (often referred to as doctorate or master's degrees), although a search using a web browser reveals many others. Professional schools offer degrees or credentials associated with a particular profession, such as law (LL.B), medicine (MD), business (M.B.A.), and veterinary medicine (D.V.M.), to name a few.

Online Application Form

Most schools now have online applications. Some may still require part or all of the application to be submitted on paper. Where possible, save or print a copy of the completed application form so that you have a record of your application. If you haven't learned it already during your undergraduate experience, start now: Always keep backups of everything! Usually the application will involve a lot of short answers or yes/no answers, but there will probably be areas where you need to write an essay (long or short). Take the time to think it out, write it out first, and proofread it before copying it onto the online form.

Personal Statement

These have already been discussed to some extent, but make sure that when applying to the school of your choice your personal statement is tailored to the specific program for which you are applying. Try to keep it as short as possible, while still covering all the relevant points. Applications that focus on working with a particular professor or in a particular area will tend to be taken more seriously. Researching what a professor has done and saying you'd like to work in that area always tend to make an application stand out.

Transcript

One of the worst things you can do to yourself is to order your transcript so late that you miss the deadline. It makes you appear disorganized and at worst that you simply don't care about going to the school you've applied to. Find out well ahead of time what your deadlines are and also how long it takes your current school to process transcript requests.

In addition, be aware that the school you are applying to may want more information, so have copies of course outlines or course information sheets and calendar descriptions available, or at least find out if you can download them from an online site as needed.

Special Examinations

Make sure you know if the school or program you are applying to requires scores from a special examination to complete your application. You probably already know that you need to write the MCAT (Medical College Admission Test) to apply to medical school, but different exams are needed for graduate schools, dental schools, nursing schools, and other professional schools. In particular, if you are planning to go to another country to attend school, you need to find out what the requirements are there. Some of these exams, what they are used for, and where to find more information about them are listed in Box 9.10. Since there are many of these exams for all sorts of programs, only those typically requested from most science students in North America are included.

BOX 9.10	Special Examinations Required for Application to School	
Name of Exam	**What It Is Used For**	**Where to Get More Information**
MCAT (Medical College Admission Tests)	Part of the application for all U.S. and most Canadian medical schools.	www.aamc.org/students/applying/mcat/ (This is the official website and has information for both U.S. and Canadian students.)
DAT (Dental Aptitude Test)	Used by Canadian dental schools to assess applicants. A similar test, the Dental Admission Test, is used by U.S. schools.	www.cda-adc.ca/en/dental_profession/dat/ (This site is hosted by the Canadian Dental Association.) www.ada.org/dat.aspx/ (This is the link for the American Dental Association test.)
GRE (Graduate Record Examination)	An admission requirement for many graduate schools in the United States.	www.ets.org/gre/ (This is the website for the group that owns this test.)

Continued

Name of Exam	What It Is Used For	Where to Get More Information
OAT (Optical Admission Test)	An admission requirement for all schools of optometry in the United States. No information was found for the two schools in Canada.	https://www.ada.org/oat/index.html (This is the website for the group that owns this test.)
PCAT (Pharmacy College Admission Test)	An admission requirement for pharmacy schools in North America.	pcatweb.info/ (This is the website for the group that owns this test.)
VCAT (Veterinary College Admission Test)	No longer used. Most veterinary schools now require MCAT or have no particular admission exam requirement.	www.aavmc.org/Publications/VMSAR.aspx (This is a link to a publication regarding veterinary school admission requirements in the United States and Canada.)
SAT Reasoning Test (formerly known as Scholastic Aptitude Test)	Entry to most universities in the United States (competes with ACT).	student.collegeboard.org (This is the website for the group that owns this test.)
ACT (formerly known as American College Testing)	Entry to most universities in the United States (competes with SAT).	www.actstudent.org/ (This is the website for the group that owns this test.)

Many examinations, such as the MCAT, DAT, and GRE, are held only at certain times each year but are offered in many different locations. For other exams, such as the NCLEX-RN (National Council Licensure Examination for Registered Nurses), there may be only one or two locations where the exam is offered, but it may be offered more frequently. Either way, you need to find out well ahead of time what exams you need to take, when and where they are offered, and how much it costs. You should also find out how to get your results to your school(s) of choice and what score you need to aim for. Last but not least, find out whether prep books or prep courses are available. Most of this information is easily available online.

The websites in Box 9.10 are a good starting place, but an online search using the name of the exam you are interested in will give you lists of resources for preparing for the exam. Be aware that you may have to start finding out these details a year or more in advance of when you actually expect to apply to the school you are interested in, especially if the exam is not offered very often.

9.11 GRANT APPLICATIONS

A grant application can be terrifying if you have never done one before, so be sure to look for examples before you start the process. Although you might find examples online, chances are that you will need to go to your campus research office or grants office and ask if they have some examples you can look at. If you know a faculty member fairly well, you can ask if you can have a look at one of her or his grant applications.

Often your supervisor will give information about where you should apply for a grant, but I know from personal experience that it doesn't always happen this way. If you don't yet know where to apply for a grant, you first need to obtain information about granting agencies in your discipline. (See Box 9.11 for a listing of some possible resources for science funding.) Again your resources would be the material you can find online, the research office, or a faculty member in your discipline. Once you have found the granting agencies, then you need to search the website of each agency for its criteria, samples of work submitted, abstracts from successful applications, and anything else that looks useful. From this you should be able to decide if the granting agency is appropriate for your proposal.

BOX 9.11   Possible Sources of Funding for Science Research in Canada

Name of Funding Agency	Contact Information
Natural Sciences and Engineering Research Council of Canada (NSERC)	www.nserc-crsng.gc.ca
Canadian Institutes of Health Research (CIHR)	www.cihr-irsc.gc.ca
Social Sciences and Humanities Research Council of Canada (SSHRC)	www.sshrc-crsh.gc.ca

Continued

Name of Funding Agency	Contact Information
National Research Council of Canada (NRC)	www.nrc-cnrc.gc.ca
Research, Technology and Innovation (Industry Canada)	www.ic.gc.ca
The Canadian Institute for Advanced Research (CIFAR)	www.cifar.ca
Canadian Innovation Centre	innovationcentre.ca
Canada Foundation for Innovation (CFI)	innovation.ca
InNOVAcorp	innovacorp.ca
International Opportunity Fund (IOF)	www.igfagcr.org/iof-home-page
Canadian Water Network (CWN)	www.cwn-rce.ca

Once you have decided to whom to apply, make sure to note any relevant deadlines and organize a schedule for yourself in order to complete your application. It will really help if you attend a grant-writing workshop; these are held at many of the major schools (so if not yours, then one nearby). You can also look online for help. Simply search "writing a grant application for funding," and examples pop up from the National Institutes for Health and other reputable agencies.

Make sure that you know the requirements of your own institution on grant applications. This will probably necessitate a visit or an email to the research office. A chat with a faculty member can be helpful to get his or her perspective on the process, but you really need to contact the research office to make sure you know all the rules and have correct information.

Many granting agencies have an ongoing grant application process. This information is available through the avenues mentioned above. Sometimes a public agency or even a private company will issue or post online a **call for proposals** for a specific type of project. Once you find a call for proposals that fits your area of research, become familiar with it; make sure you really understand what they are asking for and when the proposal is due. Then make sure you meet the deadlines.

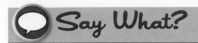

Say What?

CALL FOR PROPOSALS: An open and transparent process that permits any individual or appropriate organization to submit an application for funding to deliver services in a particular area of endeavour.

In general, proposals or grant applications require most or all of the following information:

1. The significance of the project—Will it make an original, important or novel contribution to your field?

2. The approach of the project—Describe the conceptual framework, design, methods, analyses, and assessment methods.

3. The feasibility of the project—Be sure to include

   a) a realistic time frame

   b) documented relevant experience

   c) progress to date (if ongoing project)

   d) preliminary data (if ongoing project)

   e) available resources (yours or your institution's)

   f) requested resources

   g) matching university funds

   h) sustainability of research once the funding period ends

   i) a definition of success of the project within the given budget

4. The impact of the project—How will this research impact your field?

5. Literature references

If the proposal is to be submitted by filling out an online form, be sure to print the form off or save it as an electronic document so that you can fill out a draft of the document before trying to complete it online. These forms are often complex, so you will need to take the time to proofread and restate your ideas to make the form as professional and clear as you can, while adhering to the requirements.

You will be required to determine the costs of your proposal and establish a budget, usually to a fairly detailed level. By detailed, I am not suggesting you need to list how many pens you will use, but rather you will need to include an estimate for the costs of office supplies, chemicals, equipment that needs to be purchased, and so on. The specifics will vary depending upon the project.

You should also make sure to include any collaborators and ensure that they see a copy of your document before you submit it. Once you have made a draft of the entire proposal, have had your collaborators read through it, and have had someone proofread it, you can finally begin the process of typing up the proposal and submitting it. It is best to read it again at least once in detail before hitting that submit button. As always, be aware of your spelling and grammar. Nothing ruins a proposal as quickly as bad spelling.

9.12 REFEREES

The process of getting referees or letters of reference should start well ahead of the deadline. The part where you ask someone to be a referee or to write a reference for you is not hard, although it can be hard on your ego if they say no. The hard part is selecting someone who knows you well enough to give a good referral and not just some generalized statements. Please don't ask them whether they will give you a good referral. If you don't already know the referee well enough to know how they will answer that question, then you likely shouldn't be asking them to write you a reference letter. This is one area where getting to know your professors on a friendly but respectful level can be of great value. Professors are human; if you are difficult and antagonistic toward them, it is likely they will (indeed some would say that they must) mention this in their reference letter. Whereas if you have participated positively in class, mentored your peers, and done other things to illustrate the qualities that will make you an asset to the program you are applying for, they will mention that. With regard to graduate school, it is absolutely crucial to have your honour's thesis supervisor provide a reference. When this doesn't happen, warning bells go off in the mind of whoever is reviewing your application.

Once the individual has agreed to be a referee or write a letter for you, ask them what information they would like from you. You can offer to send them an email with a document where you describe why you want to be whatever it is you are planning on. Do everything you can to give them the tools to write you a good letter. Make sure to give your referee clear deadlines and stamped self-addressed envelopes (or the offer of such, which they can then refuse).

If the school for which you want the letter also requires a form to be filled out by your referee, read it over first and see if you need to sign anything. Many schools have a form titled "Letter of Recommendation" or something similar, which first requires that you fill in the top portion and waive your right to review the letter before anything else is written on it.

Once you have asked and the individual has agreed to write the letter, make sure to follow up. I know of more than one candidate that had their application rejected simply because one of their letters of reference did not arrive on time. Most of your professors are pretty busy, and while they sincerely plan to write your letter, they may get distracted by the many other things they need to do. If all you need is permission to use their name, then just ask your professor and move forward, but if you need a letter you need to allow lead time and then follow up. Gentle and courteous nagging can be appropriate, because faculty members are often extremely busy. Most people appreciate a friendly reminder. If this concerns you, ask them if they are OK with you checking up on the status of the letter by email. However, do note that your friendly reminder will be more useful if it is received by your referee a few days before the letter is due, rather than five minutes before on the day of.

Last but not least, send your referee a thank-you letter after your application has been processed; later, when you know the outcome, let her or him know how things went. If your professor was interested enough to write you a letter, he or she will probably be interested to know how things went.

9.13 THE INTERVIEW

If you've made it to the interview stage, great! This means they want to know more about you and you have your chance to shine relative to the other candidates. Although an interview (whether for a job or a position in graduate school) can be stressful, you can best reduce your stress by being as prepared as possible.

One of the major questions you can ask yourself is, "Why do I want this particular job?" or "Why do I want to work with this particular professor?" or "Why do I want to go to this school?" I recognize that in some cases you are just desperate for a job and so anything that will pay you money will do, but what I am trying to focus on here are the jobs, research positions, and schools that will actually move your career plans forward.

So, first of all, what exactly are your career plans? This is an easy question for some but really hard for other people, and I wouldn't be at all surprised if you answered "I don't know" at this point. So now would be as good a time as any to sit down and have a think about it! If you are planning a career based on your parents' plans for you, then you need to decide if this is also what you want. If a lot of different things interest you, perhaps you need to decide which you like best or investigate finding a career in a multidisciplinary area such as the environment (only one example; there are many others). Whatever you end up deciding to do, you will be much more convincing in an interview if you have actually thought about why you are doing this.

There are different types of interviews and interview styles. It may be a one-on-one interview, or you may be interviewed by a committee. The style and setting may be formal or informal.

Many online resources of standard questions for any possible interview scenario can be found using a search engine such as Google. A few really common questions are listed below in Box 9.12. Review these and then prepare sample answers. You will probably be asked completely different questions at the interview, but the time taken to prepare will not have been wasted.

BOX 9.12   Sample Interview Questions

Can you tell us something about yourself that wasn't included in your application? What strengths will you bring to this position?

Describe the most rewarding situation that you have experienced in your career, and explain what made it particularly rewarding.

Describe a professional situation in which you had to deal with a particularly difficult person. What was it about that person that made him or her difficult to deal with? How did you handle the situation?

Continued

What do you regard as the biggest challenge facing employees in [the job you are applying for] today?

Can you tell us what you would consider to be an appropriate wage?

Can you describe a new [appropriate to the situation] technique that you have tried or would be willing to try?

Can you please elaborate on your work ethic?

Can you think of situations where you have demonstrated initiative in the past?

What is your experience throughout your career in the service/leadership area?

Could you tell us what you know about [the company you are applying to] and how this position would fit in with your career plans?

Is there anything else you would like to tell us or discuss with us?

Make sure you research the company or school with whom you are interviewing. You need to have knowledgeable answers that specifically relate to that company or school. Be honest in your answers; lies will often trip you up. If you are caught lying during or after the interview, it will usually cost you your job or position in school. If your honest answer is going to hurt your chances at the position, you will have a moral dilemma that you should resolve before you go to the interview. If the problem is that you don't know the answer to the question it's better to think about things, and then say "I don't know" and discuss why you are unsure about an answer. It shows that you think about things rather than just giving up. Prepare for the worst, most difficult question you can imagine them asking you—that way you will probably be pleasantly surprised at how well things go.

Interviews can be draining for both you and the people doing the interviewing. If the interviewers have done several interviews before yours, they might be tired or bored. In this case, don't let their demeanour influence how you perform in the interview—you still need to be positive. Make sure you really pay attention during the interview! No one is impressed if you have to keep asking the committee to repeat the question or if you give an answer that has little or nothing to do with the question that was asked. Try to speak in a tone of voice that suggests you are pleased to be there. Speak clearly, enunciate, and give answers that are short and to the point without being just yes or no. You have to walk a fine line between saying too much (far too common) and not saying enough (which doesn't give the interviewer much to work with).

If you really haven't got a clue what you are being asked, you could ask the person to rephrase the question, but try not to do this more than once. It is far better for you to say something like "I think you are asking me …," where you restate the question and then give your answer. You might end with "Did this answer your question?" and be prepared to try again if you didn't get it right. The committee needs to feel that you are intelligent and paying attention. They also need to know that you have good language skills, or they will wonder who wrote your resumé for you!

Speaking of the interview and your resumé: If you have an interview involving a committee, it is not a good idea to assume everyone has read your resumé. In a perfect world they would have, but the world is far from ideal, so often at least one person won't have read it thoroughly. You may be asked questions that were answered in your application, but it is not helpful to your chances of success to say, "Well, the answer to that question is on my resumé!" Don't let yourself get rattled by someone asking a question that you feel you have already answered either in the application or, sometimes, earlier in the interview.

Before leaving this section, let's talk briefly about interview etiquette. This includes how you dress and how you behave in the interview. We all likely know people who didn't get positions because they dressed unconventionally—let's face it, you're saying "I don't fit in" when you do that. The potential employer will wonder whether they can work with you. If you're incredibly articulate and likable the interviewer might be able to get past this, but not everyone is, so don't plan on being able to win them over in spite of how you dress. The best idea is well groomed and business-like but not flashy.

Having good manners is essential, whether at a job interview, at dinner, or when visiting a foreign country (see Box 9.13). Attention to detail makes you look good and can make a huge difference to the interview.

BOX 9.13   Good Manners and the Job Interview

Before your interview, make sure you are prepared.

1. Review your online presence. Make sure you remove any unprofessional or embarrassing posts and photos, or adjust your privacy settings to ensure a prospective employer can't access them. Make sure the email name you use to correspond with employers is business-like (not, for example, crazykid@hotmail.com).

Continued

2. Assemble the materials you will bring to the interview: reference information, your resumé, a list of questions to ask, a notebook, business cards if you have them, and a professional-looking briefcase or binder. (Avoid lugging a backpack.)

3. Dress professionally and conservatively; make sure your clothes are clean and tidy and your shoes are polished. Practise good grooming and hygiene.

4. Research the route to the interview location and how long it will take you to travel there. Plan to arrive early so you are certain that you'll be on time if there are unexpected delays; this also gives you time to freshen up before the interview. However, take care not to arrive too early—this may disrupt the interviewer's schedule.

When you arrive at the interview location, focus on creating a good first impression.

5. Dispose of any gum, food, or drink before you enter the reception area.

6. Turn off your cell phone and any watch or pager alarms.

7. Greet the receptionist in a warm, friendly manner and state your name, the purpose of your visit, and the name of the person you are meeting for the interview. Be cooperative if there are any security measures such as a visitor's sign-in or badge. Rudeness to the receptionist has been known to cost promising interview candidates the job!

8. Wait patiently for the interviewer to arrive. Read any institutional literature in the reception area to demonstrate an interest in the organization.

9. Greet the interviewer in a warm, friendly manner and again state your name and the position for which you are interviewing. Offer a firm handshake. Repeat the interviewer's name if you are hearing it for the first time and take care to get the correct pronunciation. During the interview, address the interviewer as Dr., Ms., or Mr.

10. When you arrive at the interview room, do not sit down until you are invited to do so or until the interviewer gestures that you should take a seat.

11. If you are being interviewed by a committee, pay attention to the introductions. Make eye contact with each member of the committee, shake their hands, and repeat their names.

12. At the end of the introductions you may hand out your business card, if you have one, or a copy of your resumé. Often the interviewer or committee members will give you a business card in return. Save the business cards carefully; do not stuff them in your pocket.

Continued

During the interview, be conscious of your non-verbal as well as your verbal communications.

13. Make eye contact with the interviewer when you answer a question. If you are being interviewed by committee, make eye contact with the various members as you answer.

14. Smile and maintain a relaxed, unstressed facial expression to convey that you are at ease with the interview process.

15. Sit straight and use good posture to convey respect and attentiveness. Don't fidget, smooth your hair, or adjust your clothes.

16. Respect the interviewer's personal space. Don't drop your binder or purse on the interviewer's desk; don't comment on personal items in the office.

17. Don't rush to answer a question. Most interviewers expect you to pause and think about your answer. If you don't understand the question, rephrase it and ask the interviewer if that is what is being asked, or request that the interviewer repeat the question.

18. Speak clearly and enthusiastically. Vary your tone to avoid a drone or monotone. Use correct grammar and avoid slangy expressions.

19. Pay attention to the interviewer's non-verbal cues that he or she has understood your answer. Don't go on and on in your response.

20. Be professional. Don't criticize previous jobs or employers; this suggests that you may be equally negative about your prospective employer should things not work out. Even if your prior experience wasn't positive, focus on what you learned from the job.

21. Treat the interviewer's questions seriously; avoid joking or frivolous responses.

22. Be cautious when faced with questions that might cross the line of what is legally permissible. This includes questions such as, "How often did you take sick days as a student or in your previous job?" and "Will you need to make childcare arrangements?"

23. Make notes during the interview so you can follow up with any questions you might have. The interviewer will usually ask if you have questions; do not decline this opportunity. Come prepared with a few questions in advance, and don't hesitate to refer to your notes.

Your need for proper business etiquette continues even as the interview ends and afterwards.

Continued

24. Reiterate your interest in the position and remind the interviewer that you've provided full contact details on your business card or resumé. Shake hands, smile, thank the interviewer for the opportunity, and then exit promptly. Do not dawdle or hang around the interview location.

25. Send a thank-you note to the interviewer or interview committee members within a day of the interview.

26. Let your references know they may be contacted by the organization.

27. Do not comment on the interview via social media (Twitter, Facebook, etc.).

28. Follow up discreetly with the interviewer and be respectful of their time. Do not inundate the interviewer with requests to know the outcome of the interview. Most organizations will either contact you if you made it to the interview stage or inform you at the end of the interview that only the successful candidate will be contacted. If you do not get the position, thank the interviewer or person who contacts you for their time and reiterate your interest in the company or organization. Creating a positive impression may serve you well the next time a position comes up.

9.14 FOLLOW-UP

No matter what else happens, make sure to send a letter or an email after the interview, or the rejection letter, to thank the head of the committee or the hiring manager for their time in looking at your application (see Box 9.14). This shows that you have some manners and allows you to let them know that you are interested in being considered if any future openings arise.

Box 9.14 Sample Thank-You Letter

15 Waterloo Road
Metropolitan, Saskatchewan
S0S 0S0

1 August 2012

Mr. Walter Langorski
Human Resources Manager
Astrobe Laboratories Ltd.
1632 Bayonne Street

Continued

Two Planes, Ontario
M0M 0M0

Dear Mr. Langorski:

Thank you very much for the opportunity to interview for the position of Junior Technician with Astrobe Laboratories Ltd. It was really nice to meet you and your colleagues, and I particularly enjoyed the tour of the facility.

Since the interview I have been even more excited about the possibility of working for your company, and I hope that you will consider my application package and interview in a positive manner.

I would like to thank you once again for the opportunity to meet with you, and I hope to hear from you soon.

Sincerely,

James Hurdsman

James Hurdsman

If, after submitting an application or having an interview, you don't hear anything back within a week or two, it is a good idea to follow up with a politely worded phone call or email. Sometimes people just get busy and forget to call you back; other times they may be waiting for another individual to accept or reject their offer but they are still keeping you and your application top of mind. Unfortunately, a lot of employers neglect to inform unsuccessful candidates. It's perfectly all right to look out for your own interests with a quick, polite call inquiring about your status as a candidate.

9.15 CONCLUSION

None of what is written in this chapter guarantees that you will get the position or grant that you are applying for, but it will at least allow you to have a competitive advantage by being a little better prepared. I hope you have been able to consider the things you can do to make you and your qualifications stand out from the crowd. The last thing to say is that if you are sure of your path forward, then don't give up no matter how long it takes. I had a colleague in school years ago who wanted to be a doctor and applied to various medical schools for eight years while completing a B Sc. and then an M.Sc. I lost track of him until many years later, when I walked into a hospital room with my very sick child and found this man was our physician. He was a fantastic doctor, and I was incredibly glad that he had never given up on his goals even after years of being rejected. You simply can't tell what the future holds, but you can certainly affect that future by putting your best foot forward.

FURTHER READING

Azuma, R. T. (2003). *A graduate school survival guide.* http://www.cs.unc.edu/~azuma/hitch4.html.

Camenson, B. (2003). *Great jobs for biology majors.* Toronto, ON: McGraw-Hill.

David, H. B., Tyson, J. F., & Pechenik, J. A. (2009). *A short guide to writing about chemistry.* Toronto, ON: Pearson Education.

Doyle, A. (n.d.). *Resume or curriculum vitae (CV)?* http://jobsearch.about.com/od/curriculumvitae/a/curriculumvitae.htm.

Pechenik, J. A. (2012). *A short guide to writing about biology* (8th ed). Toronto, ON: Pearson Education.

Rigden, J. S. (2003). *Landing your first job: A guide for physics students.* Melville, NY: American Institute of Physics.

Rowh, M. (2005). *Great jobs for chemistry majors* (2nd ed.). Toronto, ON: McGraw-Hill.

Swartz, M. (2012). *What's the difference between a resume and a CV in Canada?* http://career-advice.monster.ca/resumes-cover-letters/resume-writing-tips/whats-the-difference-between-a-resume-and-a-cv-in-canada/article.aspx.

10

Conclusion

"You have brains in your head. You have feet in your shoes. You can steer yourself any direction you choose. You're on your own. And you know what you know. And YOU are the one who'll decide where to go. …"

DR. SEUSS, *Oh, The Places You'll Go!*

10.1 FINAL THOUGHTS

10.2 TEN TAKE-HOME MESSAGES

10.3 SUMMARY

10

10.1 FINAL THOUGHTS

The purpose of this book is to help you develop effective scientific thinking, studying, and communication skills, and to better prepare you for your university studies and future career. Now that you've finished the book, I encourage you to keep it close at hand. In fact, I hope that you will have a long relationship with it. For example, you can consult specific sections and tables when you need a quick answer, or you can consult the word root tables in Chapter 4 and Appendix B when presented with a new term you don't know. The abbreviation tables in Chapter 6 and in Appendix B will help you when you see a new acronym but don't know its meaning. Remember, this book is designed not only to give you a foundation for studying in the sciences but also to act as a go-to guide for future questions you may have as you take new courses.

While it's tempting to think that you'll leave all your first-year books behind once you move on to your later academic years, in practice introductory textbooks can be great resources for the later years of your degree. During second year, you'll be challenged to build on concepts you met in first year, and having your introductory books at hand will help you quickly brush up on foundational ideas. Even in your final year, and perhaps beyond, you may still need to look up the details of a concept you first saw four years ago. While the money you get from selling a used book is nice, it's a small and short-term benefit. Experienced scientists will tell you that your university texts are great resources for later on in your career.

Furthermore, I hope you've absorbed one of the key messages of the book, namely that the splitting of "science" into biology, chemistry, and physics is merely a useful tool. The separation is a highly artificial one that we use to make things simpler for ourselves. While each discipline may contain its own terminology and distinct experimental methodology, they all use SI units of measure and rely heavily upon disciplined thinking about cause and effect. As interdisciplinary research grows in popularity, the boundaries between the fields are becoming increasingly blurred, so it's a good idea to be flexible in how you think about the sciences.

In addition, I encourage you to access this book's website, science3.nelson.com, to learn more about different study techniques, games, and tools that will help you learn key science concepts and science process skills. The website is a hub of science information and learning strategies. It is your go-to place to get more information on all of the topics presented in this book.

10.2 TEN TAKE-HOME MESSAGES

I'd now like to recap what I consider to be the top 10 take-home messages of this book.

1. What Is Science?

   - Science is a way of learning about—and knowing about—the natural world around us. Remember that science is not a list of facts to be remembered. In fact, if you look at university science education programs, you will see that the emphasis of science education is actually on skill development rather than content regurgitation.

2. Thinking Like a Scientist

   - You have to be focused, pay attention, and question everything. You have to be prepared to change your views when evidence arises that contradicts what you thought you understood. Keep an open mind. The first element of thinking like a scientist is to be curious and ask questions. Most professors welcome questions during lectures, so ask away.

3. Proper Preparation

   - You cannot cram for science courses and retain information or do well on exams. You need to learn new scientific knowledge, and then apply that knowledge through practice, to do well. If you're taking an exam and all you have are facts swirling around in your brain, you probably won't be able to connect ideas or evaluate your test questions at a higher level than just regurgitation. Plan your time carefully so that you actually understand the concepts. Don't procrastinate! It will probably just make you panic and make it harder to show how well you understand the ideas.

4. Focus on Learning Word Roots

   - Different scientific disciplines use different terms. Focusing on learning word roots will lead to a greater and deeper understanding of these terms. Furthermore, you'll be able to figure out new terms when you're exposed to them for the first time.

5. Active Learning Works

   - Incorporating active learning strategies into your study routine (and actively participating in class exercises when your instructor includes them in lecture) will lead to greater understanding of the subject material.

6. Test Yourself

   - Check your understanding as you go along, and routinely evaluate your understanding. Remember from Chapter 7 that tests increase learning and retrieval of information. Tests should not be the beginning or end of your study strategy—they should form the middle, wherein you test what you've been studying, and then take some time to correct the mistakes that were identified through the test.

7. Don't Be Afraid of Failure

   - Listen to feedback from your professor and TA, and apply that feedback on your next assignment or test. When you're told your answer is wrong or insufficient, don't get defensive: consider what a better answer would be, and look for habits you might have which prevented you from supplying an answer more indicative of "A" (superior) work.

8. Strive for True Understanding

   - There's a big difference between knowing the name of something and actually understanding a concept that underlies the name. You could know the name of a concept in 10 different languages, but at the end of the day you'd be less well off than knowing what it actually means in a single language. This is crucial: What you understand in one course will often be needed as the foundation for a subsequent course. Keep your notes and resources at hand. You never know when they might come in handy!

9. Keep the Conversation Going

   - At the beginning of this book, I said that I view my classes as a conversation between my students and me. I encourage you to keep the conversation going. Go to your professor during office hours and talk to him or her about learning strategies and course content. Talk to your fellow students. Talk to your TAs. Talk to anyone involved in your course. So long as you have thoughtful questions to ask or insightful points to make, you won't be wasting their time or your own!

10. Take Ownership of Your Learning

   - The only person who is responsible for your success is *you*. Take learning seriously, and take responsibility for your own study habits and learning strategies.

10.3 SUMMARY

In summary, no matter what techniques or strategies you decide to use from this book, remember that it won't be easy to get high grades in your science courses. It will require a lot of hard work. However if you employ some of the strategies we've discussed, then you'll likely study more effectively and will achieve more. I like to think of this as studying *smart* rather than studying *hard*. Just as an athlete doesn't become a gold-medal winner overnight, you can't learn all the material in your course in a few evenings at the end of term. You need to gradually develop your science knowledge and skill base. In Chapter 9 (Box 9.1), we explored transferable skills—skills you'll build upon throughout your university education and your later career. The thing to remember about transferable skills is that you'll need them later in life. Yes—there *is* life beyond university, and that is where you'll need to apply what you've learned to be successful. I wish you every success in your journey through science and your science courses!

> *"You're off to Great Places!*
>
> *Today is your day!*
>
> *Your mountain is waiting,*
>
> *So … get on your way!"*
>
> DR. SEUSS, *Oh, The Places You'll Go!*

An Overview of Scientific Literature*

This appendix is geared toward introducing the fundamental terms and concepts needed to understand and search for information in the sciences. Each of the key terms and concepts is described and defined to ensure that you have a clear grasp of what constitutes the scientific literature.

Peer review refers to a process used by journals to prevent the publication of irrelevant experiments, unsupported interpretations, bad science, and biased opinions. When an article is submitted to a peer-reviewed journal, the editors locate peers (i.e., other scientists in a similar or related field) and have them critique the article based on a range of criteria, including the writing itself, the methodology used, the assumptions made, the validity of the hypothesis, the appropriateness of the experimental design, the relevance of the research to the field of study, the conclusions drawn, and the interpretations made. Peer review may also be called **refereeing**, as the peers are expected to be neutral and have authority within the field, and their judgment is used to make a final "call" or decision. These peers recommend whether or not an article is acceptable for publication. Depending on the areas of weakness identified, the author is given the opportunity to fix/address the problems and resubmit the article for further review.

A **scholarly article** is an article that is published in a peer-reviewed journal (also called a **scholarly journal**). These articles may be written by one person or by many people. A scholarly article is intended for a professional or academic reader, and the language in which it is written tends to be formal and scientific (i.e., lots of hard-to-understand terminology!). Be aware that not all items in a scholarly journal are scholarly articles. A peer-reviewed journal will also publish opinions (also called editorials), news items, letters to the editor, and book reviews. Some of these other items might include a reference or two but a scholarly article ALWAYS includes a list of references.

*Appendix A: An Overview of Scientific Literature prepared by Mindy Thuna, University of Toronto at Mississauga.

Another type of publication that you might find is a **magazine** or **newspaper article** (also called a **popular article**, as it is written in language that is easily understood by the majority of the population). Unless your assignment indicates that you can use these as sources, they are best avoided because they are neither peer-reviewed nor original research.

Another set of terms that you need to be aware of are **primary**, **secondary**, and **tertiary sources**. These terms define a resource based on how close to the original research the information is. Thus, a **primary source** comes first and contains original research on which other research is based. This is usually the first formal appearance of experimental results and MUST contain a results section and be written in the first person, e.g., "I (or we) conducted this experiment." A **secondary source** comes next and describes, interprets, analyzes, evaluates, comments on, and discusses the content found in the primary source. In other words, the content of a secondary source repackages and reorganizes the already-published information. The final level is a **tertiary source**, which compiles, analyzes, and condenses secondary sources into a convenient, easy-to-read form (like a textbook).

CITATION AND CITATION MANAGEMENT

A **citation** is a reference to a specific work (e.g., a scholarly article) or portion of a work (e.g., a book chapter). A citation is composed of different pieces of information depending on what type of material is being cited. Therefore, if you know how to read a citation and an item is cited correctly, you can tell immediately if you are looking for a journal, a book, a magazine, or a different source. To confuse things further, there are different citation styles. You have probably heard of some of them—MLA, APA, Chicago, ACS, CSE (to name a few). The most important thing to remember is that the style is just the way a journal or organization has decided they want the information in the citation to look—i.e., the order and formatting of the information—but MOST of the information you need should be there regardless of the style used. In the sections below I have used APA format to cite some common examples.

Books

A citation to a book always includes publisher information. Because journals, magazines and newspapers do not include this information, books are relatively easy to identify.

A typical book citation looks like this:

Author. (Year of Publication). *Book Title*. Location of Publisher: Publisher.

Karl, T. R., Melillo, J. M., & Peterson, T. C. (2009). *Global climate change impacts in the United States*. New York, NY: Cambridge University Press.

The one tricky thing about a book is that sometimes you will have a citation that refers to a particular chapter in a book. This means you will have more than one title—the book title and the chapter title, BUT, unlike with journals, magazines, and newspapers, the publisher information is still included at the end.

Here is an example of what I mean:

Chun, H., David, D. K., & Zaqistowki, J. (2004). The chemistry of antioxidant constituents of *Echinacea*. In S. C. Miller & H.-C. Hu (Eds.), *Echinacea: The genus Echinacea* (pp. 73–85). Boca Raton, FL: CRC Press.

As you can see, there is a lot of extra information that is not in a "typical" book citation.

Journal Articles

There are two key pieces to a journal article citation—one is that you have two titles (one for the article and one for the journal), and the other is that you have a volume, sometimes an issue, and then page numbers for the article.

A typical journal citation looks like this (note that retrieval dates are necessary only for items that may change over time, such as Wikis):

Author. (Year). Article Title. *Journal Title, Volume*(Issue), Pages.

Vörösmarty, C. J., Green, P., Salisbury, J., & Lammers, R. (2000). Global water resources: Vulnerability from climate change and population growth. *Science, 289*, 284–288.

Webpages

A typical webpage citation looks like this:

Author. (Year of Publication). Page Title. Retrieved from *http://url*

Convention on Biological Diversity. (2004). Text of the Convention on Biological Diversity. Retrieved from *http://www.biodiv.org/convention/articles.asp*

CITATION MANAGEMENT TOOLS

Citation management tools allow you to organize your citations; create bibliographies in many citation styles; and import references from article databases, library catalogues and web sources. Some of the parameters for you to consider when choosing which citation management tool to use are

- cost
- operating system compatibility
- word processor integration

Go to the *Science*[3] website at science3.nelson.com for a list of some of the most popular citation management tools and more information on these parameters for each of them.

SEARCHING (AND FINDING) THE LITERATURE

Keywords are descriptive words used when you search for information on any topic. Typically, your results will include these words in the title, abstract, or subject heading/descriptor, depending on where you conduct your search. These words can be general or specific, depending on the information that you are trying to find.

A **database** is a collection of data that have been systematically organized. Universities and colleges purchase access to databases that include citations to scholarly journal articles to help you find these articles more easily using keywords relating to your topic.

APP A

Science³ WorkBench
Key Tables and Tools to Use

CHAPTER 2 WHAT IS SCIENCE?

Table 2.1 Canadian Salary Data from the 2006 Census

Biology	$64 350*
Chemistry	$65 700
Physics	$81 800

Remember that these salaries are national averages for people who have undergraduate degrees in the specific field, which means the numbers aren't for people who've just graduated. Finally, Statistics Canada data show that for all fields of study, a graduate degree increases your salary by over $10 000.

Data source: Statistics Canada, 2006 Census.

CHAPTER 3 THINKING LIKE A SCIENTIST

Table 3.1 Good and Bad Hypotheses

Discipline	Good Hypothesis	Bad Hypothesis
🧬 Biology	Because this synthetic molecule is similar in shape to the natural hormone testosterone, we propose it can serve medically to compensate for patents with a naturally low level of testosterone.	Because this plant extract is naturally occurring, it won't have harmful side effects if you take it as a medicine.

Continued

Discipline	Good Hypothesis	Bad Hypothesis
Chemistry	Because molecule X has the same absorption pattern in the ultraviolet range as molecule Y, at least a portion of their structures must be identical.	Because molecule X has the same absorption pattern in the ultraviolet range as molecule Y, they must have identical structures.
Physics	The vector law of adding relative velocities applies only for speeds typically measured for projectiles in terrestrial laboratories.	Because the Sun glows in the sky, its primary power source is light.

CHAPTER 4 UNDERSTANDING SCIENCE

Table 4.1 The Greek Alphabet

Greek Alphabet Uppercase Letter	Greek Alphabet Lowercase Letter	English Spelling of Greek Letter
A	α	alpha
B	β	beta
Γ	γ	gamma
Δ	δ	delta
E	ε	epsilon
Z	ζ	zeta
H	η	eta
Θ	θ	theta
I	ι	iota
K	κ	kappa
Λ	λ	lambda
M	μ	mu
N	ν	nu

Continued

Greek Alphabet Uppercase Letter	Greek Alphabet Lowercase Letter	English Spelling of Greek Letter
Ξ	ξ	xi
O	o	omicron
Π	π	pi
P	ρ	rho
Σ	σ	sigma
T	τ	tau
Y	υ	upsilon
Φ	φ	phi
X	χ	chi
Ψ	ψ	psi
Ω	ω	omega

Table 4.2 Common Symbols Used in Science

Symbol	Meaning
Å	angstrom
f or v	frequency
∞	infinity
Ø	empty set
℧	mho
∝	proportional to
⊙	Sun
⊕	Earth

Continued

Symbol	Meaning
[ ]	concentration
!	factorial
Δ	change or heat
♂	male
♀	female
°	degree

Table 4.3 Greek and Latin Numbers Used in Science

Number	Greek Prefix	Latin Prefix	Common Example
1	mono-	uni-	monogamy, uniform
2	di-/dy-	bi-/duo-	diurnal, biannual
3	tri-	tri-/terti-	triathlon, tertiary
4	tetra-	quadri-/quart-	tetrad, quadruple
5	penta-	quint-	pentagon, quintuple
6	hex-	sex(t)-, se-	hexagon, sextet
7	hepta-	sept-	heptanol, September
8	octa-	oct-	October
9	ennea-	non-/novem-	nonane, November
10	deca-	dec(a)-	December

Source: D.M. Ayers (1972). *Bioscientific terminology: Words from Latin and Greek stems.* Tucson, AZ: The University of Arizona Press; T.F. Hoad (Ed.). (1996). *Oxford concise dictionary of English etymology.* New York, NY: Oxford University Press.

Table 4.4 Common Greek and Latin Roots, Prefixes, and Suffixes in Science Terminology

Root, Prefix, or Suffix	Meaning(s)	Informal Example*	Science Example	Source†
a(n)-	without, not	atheist	anemic, asexual reproduction, anion	Greek a(n)-, meaning "not"
acro-	terminal, end, summit, tip	acronym	acroscopic, acrospore	Greek akros, meaning "terminal"
ad-	toward	address	adductor, adapt	Latin ad, meaning "toward"
aden(o)-	gland		adenoid	Greek aden, meaning "gland"
aero-	air, gas	aerodynamic	aerate, aerobic respiration	Greek aer, meaning "air"
agglutinat-	to glue together		agglutinate, agglutination	Latin agglutinare, meaning "to glue"
allo-	other, different	allophone	allogamy, alloplasty, allopatric speciation	Greek allos, meaning "other"
alveol-	hollow, cavity		alveolus	Latin alveus, meaning "concave vessel"
amphi-	around, both, two, dual	amphitheatre	amphipathic	Greek amphi, meaning "on both sides"
ana-	up, again, anew	analysis	anatomy, anabolic	Greek ana, meaning "up"
andr(o)-	male, man	philander	androgen, androgynous	Greek andro, meaning "man"
angi-	seed vessel		angiostomatous, gamentangium	Greek aggeion, meaning "vessel"
ann-	yearly, year	annual	perennial	Latin annus, meaning "year"

Continued

Root, Prefix, or Suffix	Meaning(s)	Informal Example*	Science Example	Source†
ante-	before	antecedent	antenatal	Latin *ante*, meaning "before"
anth(o)-	flower	anthology, chrysanthemum	anther, exanthema	Greek *anthos*, meaning "flower"
anthrop(o)-	man, human being	anthropology, philanthropy	anthropophilic	Greek *anthropos*, meaning "man"
anti-	against, opposite, instead of	antitrust	anticodon, antibody	Greek *anti*, meaning "against"
apic-	top, tip, summit, apex	apex	periapical, apical meristem	Latin *apic-*, the stem of *apex*, meaning "top"
apo-	away, off	apostrophe	apoptosis	Greek *apo*, meaning "away from, off"
aqu-	water	aqueduct	aquifer, aqueous humour	Latin *aqua*, meaning "water"
arachn-	spider	arachniphobia	Arachnida, arachnidium	Greek *arakhne*, meaning "spider"
arch(e)-	ancient, primitive, beginning	archaeology, archetype	adrenarche, archaea	Greek *arch-*, meaning "first, beginning"
arthr(o)-	jointed, articulated	arthritis	arthropod	Greek *arthron*, meaning "joint"
asc-	bag, sack		ascus, Ascomycetes	Greek *askos*, meaning "sack"
astr-, aster-	star	astrology, disaster	asteroid, Asteroidea	Greek *aster*, meaning "star"
atri-	room	atrium	atrium, sinoatrial	Latin *atrium*, meaning "court, hall, room"

Continued

Root, Prefix, or Suffix	Meaning(s)	Informal Example*	Science Example	Source†
auto-	same, self, spontaneous	autopsy, autograph, automobile	autism, autophagy	Greek *autos*, meaning "self"
aux(e)-	increase, grow, enlarge	auxiliary	auxin, auxochrome	Greek *auxein*, meaning "to increase"
bar-	weight, pressure	baritone	barometer	Greek *baros*, meaning "weight"
bene-	well	benefit	benign	Latin *bene*, meaning "well"
bi-	two	bicycle	binomial, bipolar	Latin *bi*, meaning "twice" or "two"
bio-	life, relating to living organisms	biology, biography	antibiotic, symbiosis	Greek *bios*, meaning "mode of life"
-blast	bud, sprout, germ, embryonic cell		astroblast, blastoderm	Greek *blastos*, meaning "bud, sprout"
brachi-	arm		brachiosaur	Greek *brakhion*, meaning "arm"
brachy-	short		brachysm	Greek *brakhys*, meaning "short"
bryo-	moss		bryophyte	Greek *bryon*, meaning "moss"
caps-	box	capsule	encapsulation	Latin *capsa*, meaning "box, chest"
carcino-	tumour	carcinogen	carcinoma	Greek *karkinos*, meaning "cancer"
cardio-	heart	cardiovascular, cardiac	cardioblast	Greek *kardia*, meaning "heart"

Continued

Root, Prefix, or Suffix	Meaning(s)	Informal Example*	Science Example	Source†
-carp, carp-	fruit		carpel, geocarpic	Greek *karpos*, meaning "fruit"
centr-	centre, middle	centrist	centripetal	Latin *centrum*, meaning "centre"
cephal-	head, brain		cephalopod, acrocephaly	Greek *kephale*, meaning "head"
chlor(o)-	green, chlorine	chlorine	chloroplast	Greek *khloros*, meaning "green/ yellow colour"
chondr(o)-	cartilage, granule	hypochondriac	mitochondria, chondrocytes	Greek *khondros*, meaning "granule" or "cartilage"
chord-, -chord	cord, string	cord	chordate, notochord, urochord	Greek *khorde*, meaning "cord"
chromo-, chromat(o)-, -chrome	colour	monochrome	chromosome	Greek *khroma*, meaning "colour"
chym-, chyl-	juice		chyme, parenchyma	Greek *khymos* and *khylos*, both meaning "juice"
clad-	branch		clade, cladogram	Greek *klados*, meaning "branch"
clin-	to slope, to lean		syncline, thermocline	Greek *klinein*, meaning "to slope"
coll-	glue	protocol	collagen, colloid	Greek *kolla*, meaning "glue"
co(m)-, con-	together, with	cooperate	codominance, recombinant, condensation	Latin *co(m)-* and *con-*, meaning "together"

Continued

Root, Prefix, or Suffix	Meaning(s)	Informal Example*	Science Example	Source†
contra-	against, opposite	contradictory	contraception	Latin *contra*, meaning "against" or "opposite"
coron-	crown, top	crown	coronal	Latin *corona*, meaning "crown"
cosm(o)-	universe	cosmic	cosmological	Greek *kosmos*, meaning "universe"
crani-	cranium, the skull	cranium	amphicrania	Greek *kranion*, meaning "skull"
cri-, crin-, -crine	to separate, to secrete	critic, crisis	apocrine, endocrine	Greek *krinein*, meaning "separate"
crypt(o)-	hidden	cryptic, cryptography	cryptogenic, cryptozoic	Greek *kryptos*, meaning "hidden, secret"
cut-	skin	cuticle	intracutaneous, cutin	Latin *cutis*, meaning "skin"
cyst(o)-, cysti-, -cyst	bladder, cyst, sac	cystic fibrosis	acrocyst, hematocyst, oocyst	Greek *kystis*, meaning "bladder, pouch"
cyt-	cell		cytoplasm, syncytium	Greek *kytos*, meaning "vessel"
dem-	people, country	democracy, epidemic	pandemic	Greek *demos*, meaning "common people"
dendr-	tree	rhododendron	dendron, dendrite	Greek *dendron*, meaning "tree"
derm(o)-, dermat(o), -derm	skin	taxidermy	dermatophyte, mesoderm	Greek *derma*, meaning "skin"
deuter(o)-	second	Deuteronomy	deuterium	Greek *deuteros*, meaning "second"

Continued

Root, Prefix, or Suffix	Meaning(s)	Informal Example*	Science Example	Source†
di-	twice, double	diploma, dilemma	diploid, dioecious	Greek *di-*, meaning "two"
dino-	terrible	dinosaur	dinosaur	Greek *deinos*, meaning "terrible"
dis-	apart, away, separately	discard	dispersion	Latin *dis-*, meaning "apart"
dorm-	sleep	dormitory	dormant	Latin *dormire*, meaning "to sleep"
-duct	to lead, to carry	conduct, induce	abduction	Latin *ducere*, meaning "to lead, draw"
dynam(o)-	power	dynamite, dynamic	thermodynamics	Greek *dynamis*, meaning "force or power"
ecdys-	an escape, moulting		ecdysone	Greek *ekdyein*, meaning "to take off, remove"
eco-	house	economy	ecology, ecotopic, ecosystem	Greek *oikos*, meaning "home, house"
ecto-	outside	ectopic pregnancy	ectotherm	Greek *ectos*, meaning "outside"
end(o)-	inside, within, inner	endogamy	endoplasmic reticulum	Greek *endon*, meaning "within"
enter-	intestine	gastroenterologist	parenteral	Greek *enteron*, meaning "intestine"
epi-	above, over	epilogue	epidermis	Greek *epi*, meaning "upon, at, before, near, on, over"
erg(o)-	work	energy, allergy	andrenergic	Greek *ergon*, meaning "work"

Continued

Root, Prefix, or Suffix	Meaning(s)	Informal Example*	Science Example	Source†
erythr(o)-	red		erythremia, erythrocyte	Greek *erythros*, meaning "red"
etho-	custom, habit		ethology	Greek *ethos*, meaning "custom, habit"
eu-	good, well, true		eukaryote	Greek *eu*, meaning "well"
ex(o)-	external, outside, out		exothermic, exergonic reaction	Greek *exo*, meaning "outside"
febr-	fever		febrile	Latin *febris*, meaning "fever"
ferment-	to decompose	ferment	fermentation	Latin *fermentare*, meaning "to ferment"
fibr-	fibre	fibre	fibrin, fibrinogen	Latin *fibra*, meaning "fibre"
fil-	thread	filagree	filament	Latin *filum*, meaning "thread"
fiss-	split		fission, fissure	Latin *fissus*, meaning "having been split"
flagell-	whip	flagellate	flagellum	Latin *flagellum*, meaning "whip"
flu-	flowing	flush	flux, fluid	Latin *fluere*, meaning "to flow"
gam-, -gamy, -gamous	marriage	monogamy, bigamy	gamete, syngamy	Greek *gamos*, meaning "marriage"
gastr(o)-	stomach, belly	gastronomic	gastropod	Greek *gaster*, meaning "stomach, belly"

Continued

Root, Prefix, or Suffix	Meaning(s)	Informal Example*	Science Example	Source†
gen(e)-	to produce	progenitor, degenerate	generation, genital	Latin *genus*, meaning "birth, origin"
geo-	Earth	geography	geodesy, geocentric	Greek *ge*, meaning "Earth"
glyc(o)-, gluc(o)-	sugar	glucose	glycogen, glycolysis	Greek *glykys*, meaning "sweet"
graph-, -graph, -graphy	to write	autograph, graphology	demography	Greek *graphein*, meaning "to write, scratch"
gymn(o)-	naked, uncovered	gymnasium	gymnosperm	Greek *gymnos*, meaning "naked"
hapl(o)-	single		haploid	Greek *haplous*, meaning "single"
helio-	Sun		heliocentric	Greek *helios*, meaning "Sun"
hemi-	half	hemisphere	hemibranch	Greek *hemi*, meaning "half"
hemo-	blood		hemophilia	Greek *haima*, meaning "blood"
hepat(o)-	liver	hepatitis	hepatocyte, heparin	Greek *hepar* (parallel stem *hepat-*), meaning "liver"
heter(o)-	different, other	heterogeneous	heterozygous	Greek *heteros*, meaning "the other of two, other, different"
hol(o)-	whole		hologram, holistic	Greek *holos*, meaning "whole"

Continued

Root, Prefix, or Suffix	Meaning(s)	Informal Example*	Science Example	Source†
homeo-, hom(o)-	likeness, similarity, same	homonym	homozygous	Greek *homoios*, and *homos*, meaning "same"
hydr(o)-	of water	hydrant, hydraulic	hydrophobic	Greek *hydro-*, from *hydor*, meaning "water"
hyp(o)-	below, under	hypotenuse	hypothermia	Greek *hypo*, meaning "below, under"
hyper-	above, over	hyperactive	hyperthermia	Greek *hyper*, meaning "above, over"
hyster-	later		hysteresis	Greek *husteros*, meaning "later"
infra-	below, under	infrastructure	infrared	Latin *infra*, meaning "below"
inter-	between	intersection	interspecies, interphase	Latin *inter*, meaning "between"
intra-, intro-	inside, within	introspective	intracellular	Latin *intra* and *intro*, both meaning "within"
is(o)-	equal, same	isosceles	isotonic, isotope	Greek *isos*, meaning "equal"
-it is	inflammation	arthritis	hepatitis	Greek *–itis*, meaning "pertaining to"
karyo-, caryo-	nucleus, nut, kernel		karyotype	Greek *karyon*, meaning "kernel, nut"
kine-, cine-, kinet-	to move	kinetic, cinema	kinesin, kinesiology	Greek *kinein*, meaning "to move"

Continued

Root, Prefix, or Suffix	Meaning(s)	Informal Example*	Science Example	Source†
lact(o)-	milk	lactose	lactation, lactase	Latin *lac*, meaning "milk"
lamin-	thin plate	laminate	lamella, nuclear lamina	Latin *lamina*, meaning "thin plate"
leuk(o)-, leuc(o)	white	leukemia	leukocyte	Greek *leukos*, meaning "white"
lip(o)-	fat, grease	liposuction	lipase	Greek *lipos*, meaning "fat"
lith(o)-, -lith	stone	lithography	lithosphere, megalith	Greek *lithos*, meaning "stone"
lum-	light	luminary	luminescence	Latin *lumen*, meaning "light"
-lyse, -lysis	break, loosen	analysis	lysis, hydrolysis	Greek *lysis*, meaning "dissolution"
macr(o)-	large	macroeconomics	macrogamy, macromolecules	Greek *macros*, meaning "long, large"
mal-	bad	malevolent	malignant, malpractice	Latin *malus*, meaning "bad"
mamm-	breast	mammogram	mammary gland	Latin *mamma*, meaning "breast"
meg(a)-	large	megaphone	megaspore, megalopic	Greek *megas*, meaning "great, large"
melan-	black	melancholy	melanin	Greek *melas*, meaning "black"
micr(o)-	small	microphone	microscope	Greek *micros*, meaning "small"

Continued

Root, Prefix, or Suffix	Meaning(s)	Informal Example*	Science Example	Source†
mito-	thread		mitochondria	Greek *mitos*, meaning "thread"
mon(o)-	one, single	monocle, monarch	monogeny	Greek *monos*, meaning "alone"
morph-, -morph, -morphous	form	morphine, amorphous	mesomorph	Greek *morphe*, meaning "form, shape"
mov-, mot-, mom-	move, motion	motor	momentum	Latin *movere*, meaning "to move"
mult(i)-	many	multitude	multiplex	Latin *multus*, meaning "many, much"
nephr(o)-	kidney		nephron	Greek *nephros*, meaning "kidney"
neur(o)-	nerve	neurotic	neuron	Greek *neuron*, meaning "nerve"
neutr(o)-	having no charge, neutral	neutral	neutron	Latin *neuter*, meaning "neither"
nihil-	nothing	nihilist	annihilation	Latin *nihil*, meaning "nothing"
nom-	name	nominate	binomial	Latin *nomen*, meaning "name"
noto-	back		notochord	Greek *noton*, meaning "back"
nucle(o)-	nucleus, nut	nuclear	nucleus	Latin *nucleus*, meaning "kernel, nut"
o(o)-	egg		oocyte, oogonium, oogenesis	Greek *oon*, meaning "egg"
od-	path, way		anode, diode	Greek *hodos*, meaning "a way"

Continued

Root, Prefix, or Suffix	Meaning(s)	Informal Example*	Science Example	Source†
olig(o)-	few, scant	oligopoly	oligandrous	Greek *oligos*, meaning "small, little, few"
onco-	tumour		oncology	Greek *ogkos*, meaning "bulk"
ortho-	straight, correct	orthotics	orthopedic	Greek *orthos*, meaning "straight, correct"
oste(o)-	bone	osteopath	osteoblast	Greek *osteon*, meaning "bone"
paed(o)-, ped(o)-	child, instruction	encyclopedia, pedagogical	pediatric	Greek *pais* (stem = *paid-*), meaning "child"
pale(o)-	ancient	paleography	paleobiology	Greek *palai*, meaning "long ago"
pan-	all	pantheism	pandemic	Greek *pant*, meaning "all"
par(a)-	beside, next to, near	parallel	parallax	Greek *para*, meaning "beside"
path(o)-, -path, -pathic, -pathy	disease, suffering	sympathy, pathetic	idiopathic	Greek *pathos*, meaning "feeling, suffering"
peri-	around, near	periphery	pericardium, perigee	Greek *peri*, meaning "around"
phag(o)-, -phage, -phagous, -phagy	to eat	sarcophagus	autophagy, phagocytosis	Greek *phagein*, meaning "to eat"
phen(o)-	appear		phenotype	Greek *phainein*, meaning "to show"

Continued

Root, Prefix, or Suffix	Meaning(s)	Informal Example*	Science Example	Source†
phil-, -phil, -phile, -philia, -philous	loving	philanthropy, philosophy	neutrophil, halophile	Greek *philos*, meaning "dear, beloved"
-phobe, -phobic	aversion to, fear	phobia	hydrophobic	Greek *phobos*, meaning "fear, panic"
phot(o)-	light	photograph	photosynthesis	Greek *phos* (parallel stem = *phot-*), meaning "light"
-phyll	leaf		chlorophyll	Greek *phyllon*, meaning "leaf"
phys-	nature	physical	physiology	Greek *phusis*, meaning "nature"
phyt(o)-	plant		phytoplankton	Greek *phuton*, meaning "plant"
pino-	drink		pinocytosis	Greek *pinein*, meaning "to drink"
-plast	something moulded, formed	plastic, plaster	chloroplast, plastid	Greek *plastos*, meaning "formed"
plec-	interwoven		symplectic	Greek *pleckein*, meaning "to twine, plait, weave"
pluri-	more, many	plural	pluriparity	Latin *pluri-*, from *plus*, meaning "more"
-pod, -podium, podo-	foot	tripod, podium, octopus	podiatrist, pseudopodia	Greek *pous* (stem = *pod-*), meaning "foot"

Continued

Root, Prefix, or Suffix	Meaning(s)	Informal Example*	Science Example	Source†
poly-	many	polygon	polymer	Greek *polys*, meaning "many"
-port	to carry	transport	transporter	Latin *portare*, meaning "to carry"
post-	after, behind	postscript	postsynaptic cell	Latin *post*, meaning "behind, after"
pre-, pro-	before	pretest	precambium, prokaryote, promoter	Latin *prae* and *pro*, meaning "in front, before"
prot(o)-	first, original	prototype, protagonist	protozoan	Greek *protos*, meaning "first"
pseud(o)-	false	pseudonym	pseudopod, pseudogene	Greek *pseudes*, meaning "false"
psych(o)-	spirit, soul	psyche	psychology	Greek *psykhe*, meaning "soul"
quadr-, quadru-	four	quadruple	quadruped	Latin *quadri-* and *quadru-*, meaning "four"
quant-	how much	quantity	quantifiable	Latin *quantus*, meaning "how much"
quin-	five, fifth	quintuple	quintile	Latin *quintus*, meaning "fifth"
radi-	spoke of a wheel, ray	radiator, radio	radiole, radiate	Latin *radius*, meaning "ray"
recti-	straight	rectify	rectilinear	Latin *rectus*, meaning "straight"
retro-	backward, behind	retroactive	retrograde	Latin *retro*, meaning "backward"

Continued

Root, Prefix, or Suffix	Meaning(s)	Informal Example*	Science Example	Source†
rhiz(o)-, -rhiza	root	rhizome	mycorrhiza	Greek *rhiza*, meaning "root"
sarco-	flesh	sarcophagus	osteosarcoma	Greek *sarx*, meaning "flesh"
saur(o)-, -saur	lizard		dinosaur, branchiosaur	Greek *sauros*, meaning "lizard"
-scope	observe, view	scope	telescope, microscope	Greek *skopein*, meaning "to observe"
sec-, seg-	cut	segment	secant	Latin *secare*, meaning "to cut"
semi-	half, partial	semicircle	semicircular canals	Latin *semi-*, meaning "half"
sider-	star		sidereal	Latin *sidus* (stem = sider-), meaning "star"
solv-	to loosen, dissolve	insolvent	solute, solvent	Latin *solvere*, meaning "to loosen"
som-, -some	body		dermatosome, somite, acrosome	Greek *soma*, meaning "body"
sperm-, -sperm, spermat(o)-	semen, seed	sperm	angiosperm, spermatheca	Greek *sperma*, meaning "seed"
spor(o)-	seed	sporadic, diaspora	sporont, acrospore	Greek *spora*, meaning "seed, a sowing"
sta-	to stand, to stop, to fix	static, thermostat	hemostasia, epistasis	Greek *histemi* (stem = sta-), meaning "to make stand"

Continued

Root, Prefix, or Suffix	Meaning(s)	Informal Example*	Science Example	Source†
stoma-	mouth, opening	stomach	deuterostome	Greek *stoma*, meaning "mouth"
sym-, syn-	together, with	symphony	symbiosis, synapse, sympatry	Greek *syn*, meaning "together, with"
tach-	swift	tachometer	tachycardia	Greek *tachus*, meaning "swift," and *tachos*, meaning "speed"
taxo-	to arrange, to put in order	tactics, syntax	taxonomy, taxon, phototaxis	Greek *taxis*, meaning "arrangement"
tel(e)-	afar	television, telephone	telescope	Greek *tele*, meaning "afar"
telo-	end		telomerase, telomere	Greek *telos*, meaning "end"
tetra-	four	tetrahedron	tetrapod	Greek *tettares*, meaning "four"
therm(o)-	heat	thermometer, thermostat	hypothermia	Greek *therme*, meaning "heat"
thyro-, thyreo-	shield		thyroid	Greek *thyreos*, meaning "shield"
tom-, -tomy	to cut, section	anatomy, atom	appendectomy	Greek *tome*, meaning "a cutting"
top(o)-, -tope	place	topical	isotope, epitope	Greek *topos*, meaning "place"
trans-	across, through	transport	transporter, transcription, translation	Latin *trans*, meaning "across, through"
tri-	three	triple	trimester	Greek *tri-*, meaning "three"

Continued

APP B

Root, Prefix, or Suffix	Meaning(s)	Informal Example*	Science Example	Source†
trop-	a turn	tropic	phototropic	Greek *tropos*, meaning "a turn, direction, way"
-troph	nourishment, development		heterotroph, autotroph	Greek *trophe*, meaning "food, nourishment"
ultra-	beyond	ultralight	ultrasonic	Latin *ultra*, meaning "beyond"
umbr-	shade, shadow	umbrella	penumbra	Latin *umbra*, meaning "shade"
uni-	one	unique, uniform	uniporter, uniparous	Latin *uni-*, meaning "one"
vacu-	empty	vacuum	vacuole	Latin *vacuus*, meaning "empty"
val-	to be strong, to be well	valid, valedictorian, evaluate	valence, covalent bond, bivalent	Latin *valere*, meaning "to be strong"
vas-	vessel	vase	vascular, cardiovascular	Latin *vas*, meaning "vessel"
verber-	strike, beat		reverberation	Latin *verberare*, meaning "to strike"
visc-	thick		viscosity	Latin *viscum*, meaning "a sticky substance"
viv-	living	survive, vivid, vivacious	viviparous	Latin *vivus*, meaning "living"
volv-, volut-	roll	revolve	convolution	Latin *volvere*, meaning "to turn"
vom-	discharge	vomit	vomit	Latin *vomere*, meaning "to discharge, vomit"

Continued

Root, Prefix, or Suffix	Meaning(s)	Informal Example[*]	Science Example	Source[†]
xen(o)-	foreign	xenophobia	xenobiotic	Greek *xenos*, meaning "foreign"
xyl(o)-	wood	xylophone	xylem	Greek *xylon*, meaning "wood"
zo-, -zoa(n), -zoic, -zoon	animal, living being	zodiac	metazoan	Greek *zoe*, meaning "life," and *zoon*, meaning "living being"
zyg(o)-	yoke		zygote, heterozygous, monozygotic	Greek *zygon*, meaning "yoke"

[*]*Not all roots, prefixes, and suffixes have an informal example.*
[†]*Information for these terms came from these sources:*

D.M. Ayers (1972). *Bioscientific terminology: Words from Latin and Greek stems.* Tucson, AZ: The University of Arizona Press; T.F. Hoad (Ed.). (1996). *Oxford concise dictionary of English etymology.* New York, NY: Oxford University Press.

BOX 4.1   Numerical Prefixes

1	mono-	6	hexa-
2	di-	7	hepta-
3	tri-	8	octa-
4	tetra-	9	nona-
5	penta-	10	deca-

BOX 4.2 Common Polyatomic Ions (excluding those given previously)

Name	Formula	Name	Formula	Name	Formula
acetate	$C_2H_3O_2^-$	dichromate	$Cr_2O_7^{2-}$	peroxide	O_2^{2-}
ammonium	NH_4^+	hydrogen carbonate	HCO_3^-	phosphate	PO_4^{3-}
bromate	BrO_3^-	hydroxide	OH^-	oxalate	$C_2O_4^{2-}$
carbonate	CO_3^{2-}	nitrate	NO_3^-	sulfate	SO_4^{2-}
chromate	CrO_4^{2-}	nitrite	NO_2^-	sulfite	SO_3^{2-}
cyanide	CN^-	permanganate	MnO_4^-	thiocyanate	SCN^-

BOX 4.3 Some Common Inorganic Acids (all names are followed by the word acid)

Carbonic	H_2CO_3	Hydrochloric	HCl	Nitrous	HNO_2
Chloric	$HClO_3$	Hydrocyanic	HCN	Perchloric	$HClO_4$
Chlorous	$HClO_2$	Hydroiodic	HI	Phosphoric	H_3PO_4
Hydrobromic	HBr	Nitric	HNO_3	Sulfuric	H_2SO_4

BOX 4.4 More about Oxyanions

When an element can combine with oxygen to form more than one oxyanion, use the following rules to name them. The one with the most oxygen atoms gets the -ate ending, and the other one, with fewer oxygen atoms, gets the -ite ending. Some of the halogens can combine with oxygen in four different ways. The anion with an extra oxygen is named per____ate. The anion with the fewest oxygens is named hypo___ite. For chlorine, the series is perchlorate (ClO_4^-), chlorate (ClO_3^-), chlorite (ClO_2^-), and hypochlorite (ClO^-).

BOX 4.5   Development of Ideas for Naming

Alkane − H = alkyl group

Molecular Formula	Alkane Structure	Alkane Name	− H	=	Alkyl Group Structure	Alkyl Group Name
CH_4		Methane	− H	=	CH_3—	Methyl
C_2H_6	CH_3CH_3	Ethane	− H	=	CH_3CH_2—	Ethyl
C_3H_8		Propane	− H	=	$CH_3CH_2CH_2$ — CH_3CHCH_3 |	Propyl *Isopropyl* (1-methylethyl)
C_4H_{10}		Butane	− H	=	$CH_3CH_2CH_2CH_2$— $CH_3CH_2CHCH_3$ |	Butyl *Sec-butyl* (1-methylpropyl)
		Isobutane 2-methyl-propane	− H	=	CH_3 | CH_3CHCH_2— CH_3 | CH_3CCH_3 |	*Isobutyl* (2-methylpropyl) *tert-butyl* (1,1-dimethylethyl)

BOX 4.6   Names of the First 10 Normal Alkanes

1 carbon—methane	6 carbons—hexane
2 carbons—ethane	7 carbons—heptane
3 carbons—propane	8 carbons—octane
4 carbons—butane	9 carbons—nonane
5 carbons—pentane	10 carbons—decane

APP B

CHAPTER 5 SCIENCE AND MATH

BOX 5.2   The Système International d'Unités (SI) Prefixes

Metric Measures and Imperial Equivalents

Unit	Measure	Symbol	Imperial Equivalent
Linear Measure			
1 kilometre	$= 1000$ metres 10^3 m	km	0.62137 mile
1 metre		m	39.37 inches
1 decimetre	$= 1/10$ metre $= 10^{-1}$ m	dm	3.937 inches
1 centimetre	$= 1/100$ metre $= 10^{-2}$ m	cm	0.3937 inch
1 millimetre	$= 1/1000$ metre $= 10^{-3}$ m	mm	Not used
1 micrometre (or micron)	$= 1/1\,000\,000$ metre $= 10^{-6}$ m	μm (or μ)	Not used
1 nanometre	$= 1/1\,000\,000\,000$ metre $= 10^{-9}$ m	nm	Not used
Measures of Capacity (for fluids and gases)			
1 litre		L	1.0567 U.S. liquid quarts
1 millilitre	$= 1/1000$ litre $=$ volume of 1 g of water at stp[*]	mL	
Measures of Volume			
1 cubic metre		m^3	
1 cubic decimetre	$= 1/1000$ cubic metre $= 1$ litre (L)	dm^3	
1 cubic centimetre	$= 1/1\,000\,000$ cubic metres $= 1$ millilitre	cm$^3 = $ mL	
1 cubic millimetre	$= 1/100\,000\,000$ cubic metres	mm^3	
Measures of Mass			
1 kilogram	$= 1000$ grams	kg	2.2046 pounds
1 gram		g	15.432 grains
1 milligram	$= 1/1000$ grams	mg	0.01 grain (about)
1 microgram	$= 1/1\,000\,000$ grams	μg (or mcg)	

[*]stp = standard temperature and pressure

Table 5.1 Outcomes from Rolling Two Dice

Sum of Dice Values (all possible x values)	Frequency of x, f(x)	Sum of These x Values, x × f(x)	P(X = x), f(x)/n	x × P(x)
2	1	2	0.01	0.02
3	6	18	0.06	0.18
4	11	44	0.11	0.44
5	12	60	0.12	0.60
6	14	84	0.14	0.84
7	16	112	0.16	1.12
8	13	104	0.13	1.04
9	10	90	0.10	0.90
10	9	90	0.09	0.90
11	6	66	0.06	0.66
12	2	24	0.02	0.24
Total	$n = 100$	694	1.00	6.94

Table 5.2 Calculating the Variance of the Two Dice

Sum of Dice Values (all possible x values)	Frequency of x, f(x)	Sum of the Squared Deviations, $f(x) \times (x_i - \bar{x})^2$	P(X = x), f(x)/n	$p(x) \times (x_i - \bar{x})^2$
2	1	24.40	0.01	0.02
3	6	93.14	0.06	0.18
4	11	95.08	0.11	0.44

Continued

Sum of Dice Values (all possible x values)	Frequency of x, $f(x)$	Sum of the Squared Deviations, $f(x) \times (x_i - \bar{x})^2$	$P(X = x)$, $f(x)/n$	$p(x) \times (x_i - \bar{x})^2$
5	12	45.16	0.12	0.60
6	14	12.37	0.14	0.84
7	16	0.06	0.16	1.12
8	13	14.61	0.13	1.04
9	10	42.44	0.10	0.90
10	9	84.27	0.09	0.90
11	6	98.90	0.06	0.66
12	2	51.21	0.02	0.24
Total	$n = 100$	561.64	1.00	5.62

BOX 5.12   Rejecting or Accepting Hypotheses—Outcomes and Possible Errors

	Action	
	Null hypothesis accepted	**Null hypothesis rejected**
Null hypothesis is actually true	Good decision	Type I error: You concluded there is a difference when there isn't.
Null hypothesis is actually false	Type II error: You concluded there was no difference even though there is.	Good decision

BOX 6.1   Table of SI Base Units

Name	Symbol	Measure	Simplified Definitions
metre	m	length	The length of the path travelled by light in vacuum during a specific time interval.
kilogram	kg	mass	The mass of a prototype platinum-iridium bar in a vault in Paris, France.
second	s	time	The time required for the light emitted by a specific energy transition of a cesium-133 atom at rest to undergo 9 192 631 770 periods.
ampere	A	electric current	The constant current that under specifically defined circumstances produces a specific force per metre of length between two conductors.
kelvin	K	thermodynamic temperature	A specified fraction of the temperature at which water of a specified isotopic composition is present in the three states—liquid, gas, and solid.
mole	mol	amount of substance	The amount of substance of a system that contains as many elementary entities as there are unbound atoms, at rest and in their ground state, in 0.012 kg of carbon 12. You must always specify "moles of what," for example, moles of protons. Typically this number is taken to be Avogadro's number of items (6.022×10^{23}).
candela	cd	luminous intensity	The luminous intensity, in a given direction, of a source that emits monochromatic radiation of a specific frequency with a specific radiant intensity in that direction.

BOX 6.2   Common and Not-So-Common Physics Abbreviations

2-D	two-dimensional
3-D	three-dimensional
A	ampere
AC	alternating current
BCS	Bardeen–Cooper–Schrieffer theory of superconductivity
BTU	British thermal unit
C	coulomb
C	heat capacity
c	speed of light, 299 792 458 m s^{-1}
CCD	charge-coupled device
CDM	cold dark matter
CFD	computational fluid dynamics
CMOS	complementary metal-oxide semiconductor
COG	centre of gravity
$\triangle d$	a small change in the variable d
DC	direct current
DE	differential equation
E	Young's modulus of elasticity
emf	electromotive force
EOS	equation of state
FEM	finite element method
FFT	fast Fourier transform
FTL	faster than light
FWHM	full width half maximum
Hz	hertz
IR	infrared
J	joule
K	kelvin
kg	kilogram
LMS	least mean square

Continued

LTE	local thermodynamic equilibrium
MRI	magnetic resonance imaging
MTBF	mean time between failures
N	newton
NIST	National Institute of Standards and Technology (U.S.)
NMR	nuclear magnetic resonance
P-P	peak to peak
Pa	pascal
PET	positron emission tomography
PDE	partial differential equation
QFT	quantum field theory
RF	radio-frequency
rms	root-mean-square
SI	Système International
SNR	supernova remnant
TEM	tunnelling electron microscopy
TNO	trans-Neptunian object
UHF	ultra high frequency
UV	ultraviolet
V	volt
W	watt
XRD	X-ray diffraction
ZPE	zero point energy

BOX 6.3 Common Physics Notation

A	area; amplitude
a	acceleration (vector quantity)
C	capacitance; heat capacity; constant of integration

Continued

c	speed of light, 299 792 458 m s^{-1}; specific heat capacity
d	distance; diameter; differential
$\mathbf{D}$	electric displacement field (vector quantity)
E	energy
$\mathbf{E}$	electric field (vector quantity)
e	base of the natural logarithm, 2.71828…; electron charge, 1.602 176 565 × 10^{-19} C
$\mathbf{F}$	force (vector quantity)
f	frequency; friction; function
G	gravitational constant, 6.67384 × 10^{-11} m^3 kg^{-1}s^{-2}
g	acceleration due to gravity at the surface of the Earth, 9.81 m s^{-1}
h	height; Planck's constant, 6.626 069 57 × 10^{-34} J s
I	current; intensity; moment of inertia
K	kinetic energy
k	Boltzmann constant, 1.380 648 8 × 10^{-23} J K^{-1}
l	length
m	mass
$\mathbf{p}$	momentum (vector quantity)
q	charge
R	resistance
r	radius
S	entropy
s	speed
T	temperature
t	time
U	potential energy

Continued

u	initial velocity (vector quantity)
V	voltage
v	velocity (vector quantity)
w	width
α	fine structure constant, $7.297\ 352\ 569 \times 10^{-3}$
β	beta particle
γ	gamma ray; Lorentz factor in special relativity
δ	often used to represent a small change in something
ε	electromotive force
η	viscosity
θ	angle
λ	wavelength
μ	coefficient of friction
ν	frequency
π	pi, $3.141\ 59\ldots$
ρ	density
σ	standard deviation; dispersion
τ	mean lifetime
ψ	wave function in quantum mechanics
ω	angular velocity

BOX 6.5 Common Chemistry Abbreviations

A	ampere
A	area
a	acceleration

Continued

C	coulomb
e	electron
F	Faraday
Hz	hertz
IR	infrared (referring to a region of the electromagnetic spectrum)
J	joule
N	newton
P	momentum; power; pressure
Pa	pascal
rf	radio-frequency
rms	root-mean-square
SI	Système International
UV	ultraviolet (referring to a region of the electromagnetic spectrum)
V	voltage; volume
V	volt
W	watt

BOX 6.6   Common Organic Chemistry Abbreviations

Ac	acetyl group
AcO	acetate
Bn	benzyl
Bu	butyl
DMSO	dimethylsulfoxide
EDTA	ethylenediamine tetra-acetic acid
Et_2O	diethyl ether
LAH	lithium aluminum hydride
Me	methyl
Ph	phenyl

Continued

n-Pr	n-propyl (the n is short for normal; normal-propyl or normal anything refers to a straight chain of carbon atoms as opposed to branched chains; it is less frequently used by chemists these days but still used in industry and other disciplines)
Pr	propyl
Py	pyridine
THF	tetrahydrofuran
THP	tetrahydropyranyl

BOX 6.7   Common Chemistry Notations*

[]	molar concentration
A	area; amplitude
a	acceleration
C	capacitance; heat capacity; constant of integration
c	speed of light; specific heat capacity; concentration
d	distance; diameter; differential
ρ or D	density
E	energy; electrical potential
e	base of the natural logarithm
F	force
f	frequency; friction; function
G	gravitational constant
g	acceleration due to gravity

Continued

h	height; Planck's constant
$H, \Delta H$	molar enthalpy (heat); a subscript is often used to denote heats of formation (f), heats of reaction (r), and other forms
I	current
J	coupling constant
KE or E_k	kinetic energy
K_{eq}	equilibrium constant; examples of constants with other subscripts are the acid ionization constant (K_a) and the solubility product (K_{sp})
l	length
m	mass
M	molar mass
mp	melting point
n	amount
PE or E_p	potential energy; photoelectric effect
Q	charge
R	gas constant
T	temperature
t	time
u	unified atomic mass unit, same value as amu
w	width
Δ	change in whatever symbol follows it; also used to denote heat when heat is added to a reaction

*Some symbols are used for more than one term, and some terms have more than one symbol in common usage. This is sometimes confusing for me, so I expect you may find it confusing too, but generally the context (or the equation you are using it in) will help you sort it out.

BOX 6.9 Basic Reaction Types Typically Seen in General Chemistry

Type of Reaction	Description	Example
Synthesis or combination reaction	$A + B \rightarrow AB$	$2Na(s) + Cl_2(g) \rightarrow 2NaCl(s)$
Decomposition reaction	$AB \rightarrow A + B$	$2HgO(s) + heat \rightarrow 2Hg(l) + O_2(g)$
Substitution or single-displacement reaction	$AC + B \rightarrow A + BC$	$Zn(s) + 2HCl(aq) \rightarrow H_2(g) + ZnCl_2(aq)$
Double-displacement reaction	$AB + CD \rightarrow AC + BD$	$Cd(s) + 2HNO_3(aq) \rightarrow Cd(NO_3)_2(aq) + H_2S(aq)$
Acid–base reaction	$HA + B \rightarrow A + HB$ or $HA + NaOH \rightarrow NaA + H_2O$	$HCN(g) + NaOH(s) \rightarrow NaCN(s) + H_2O(g)$
Redox or oxidation-reduction reaction	Always involves a loss and gain of e^-	$SnO_2(s) + 2C(s) \rightarrow Sn(l) + 2CO(g)$
Combustion	$C_xH_y + O_2(g) \rightarrow CO_2(g) + H_2O(g)$	$2C_4H_{10}(l) + 13O_2(g) \rightarrow 8CO_2(g) + 10H_2O(g)$

BOX 6.10 Common Biology Abbreviations

A	adenine; adenosine; the letter abbreviation for the amino acid alanine
Ab	antibody
Ac	acetyl group
acetyl CoA	acetyl coenzyme A
ADA	adenosine deaminase
ADH	alcohol dehydrogenase
ADP	adenosine 5'-diphosphate
Ag	antigen
AIDS	acquired immune deficiency syndrome

Continued

Ala	alanine residue (also identified as A)
AMP	adenosine 5'-monophosphate
ANOVA	analysis of variance
Arg	arginine residue (also identified as R)
Asn	asparagine residue (also identified as N)
Asp	aspartic acid residue (also identified as D)
ATP	adenosine triphosphate
ATPase	adenosine triphosphatase
β-gal	β-galactosidase
BAC	bacterial artificial chromosome
BLAST	Basic Local Alignment Research Tool
bp	base pair
C	cytosine or cytidine; also the one-letter code for the amino acid cysteine
cal	calorie (4.18 J)
cDNA	complementary deoxyribonucleic acid
CFU	colony-forming unit
CGH	comparative genome hybridization
Ci	curie
cM	centimorgan
cpm	counts per minute
cys	cysteine residue (also identified as C)
Da	dalton
DNA	deoxyribonucleic acid
DNAse	deoxyribonuclease
dNTP	deoxynucleoside triphosphate
ds	double-stranded
EGF	epidermal growth factor
ER	endoplasmic reticulum
FACS	fluorescence-activated cell sorting
FISH	fluorescence in situ hybridization
fMet	formyl-methionine

Continued

FRET	fluorescent resonant energy transfer
G	guanine or guanosine; one-letter code for the amino acid glycine
G	Gibbs free energy of a system
GFP	green fluorescent protein
Glc	glucose residue
Gln	glutamine residue (also identified as Q)
Glu	glutamic acid residue (also identified as E)
Gly	glycine residue (also identified as G)
GTF	general transcription factor
GTP	guanosine 5′-triphosphate
H	enthalpy of a system
hGH	human growth hormone
His	histidine residue (also identified as H)
HIV	human immunodeficiency virus
HLA	histocompatibility locus antigen
hsiRNA	heterochromatic short interfering RNA
HSV	herpes simplex virus
I	inosine
Ig	immunoglobulin
Ile	isoleucine residue (also identified as I)
J	joule
K	degrees kelvin (absolute temperature)
kb	kilobase
kcal	kilocalorie
K_d	dissociation constant
kDa	kilodalton
LDL	low-density lipoprotein
Leu	leucine residue (also identified as L)
Lys	lysine residue (also identified as K)
LTR	long terminal repeat
m	metre
M	relative molecular mass

Continued

mAb	monoclonal antibody
Mb	megabase; megabyte
MCS	multiple cloning site
Met	methionine residue (also identified as M)
MHC	major histocompatibility locus
mol	mole
mp	melting point
mRNA	messenger ribonucleic acid
NAD	nicotinamide adenine dinucleotide
NAD^+	nicotinamide adenine dinucleotide (oxidized)
NADH	nicotinamide adenine dinucleotide (reduced)
$NADP^+$	nicotinamide adenine dinucleotide phosphate (oxidized)
NADPH	nicotinamide adenine dinucleotide phosphate (reduced)
NCBI	National Center for Biotechnology Information (U.S.)
NCI	National Cancer Institute (U.S.)
neo	neomycin gene (used as a selectable marker)
NK	natural killer cells
NMR	nuclear magnetic resonance
nt	nucleotide
oligo	oligonucleotide
OMIM	Online Mendelian Inheritance in Man
ORF	open reading frame
ori	origin of replication
PCD	programmed cell death
PCR	polymerase chain reaction
Phe	phenylalanine residue (also identified as F)
P_i	inorganic phosphate
ppm	parts per million
Pro	proline residue (also identified as P)
Pu	purine
Py	pyrimidine

Continued

RBC	red blood cell
RBS	ribosome-binding site
RE	restriction endonuclease
RFLP	restriction-fragment-length polymorphism
RNA	ribonucleic acid
RNAse	ribonuclease
rRNA	ribosomal ribonucleic acid
RT	reverse transcriptase
RT-PCR	reverse transcriptase polymerase chain reaction
S	entropy of a system
S	svedberg sedimentation unit
siRNA	short interfering ribonucleic acid
snRNA	small nuclear ribonucleic acid
SRP	signal recognition particle
ss	single-stranded
STS	sequence tagged site
T	thymine or thymidine; one-letter code for the amino acid threonine
Taq	thermus aquaticus DNA (which is used as a polymerase)
TBT	TATA-binding protein
TCR	T cell receptor
TF	transcription factor
TGN	trans-Golgi network
Thr	threonine residue (also identified as T)
tRNA	transfer ribonucleic acid
Trp	tryptophan residue (also identified as W)
Tyr	tyrosine residue (also identified as Y)
U	uracil
UTR	untranslated region
UV	ultraviolet
Val	valine residue (also identified as V)
WBC	white blood cell
WT	wild-type

BOX 6.12   The Hierarchy of Biology

Hierarchy Level	Example	Properties Arising at This Level of Structure
Atom	Hydrogen	Electronegativity; ionization; radioactivity; and features of electron configuration (energy levels, valence) occur at this level.
Molecule	Water (H_2O)	The ability to form particular types of bonds (e.g., hydrogen, ionic, covalent, van Der Waals interactions) and the beginnings of 3-D shape occur at this level.
Macromolecule	DNA	With large molecules arising from consistent monomers, simple and repeatable molecules with specific functions are better supported.
Organelle	Nucleus	Groups of macromolecules can contribute to partitions (e.g., membranes) or form clusters of specific metabolic function (e.g., enzymes within a lysosome).
Cell	Neuron	With a complete complement of organelles, a structure that exhibits ALL the characteristics of life can arise. There is no known entity below the cellular level that demonstrates all the properties of life.
Tissue	Neural tissue	With a group of similar cells working together, new features can be achieved. A single neuron doesn't think, but a collection of neurons can form a decision-making body such as a neuron.
Organ	Brain	With the support of several cell types, an entity can be created that is capable of more complex activity, such as the formation of memories.

Continued

Hierarchy Level	Example	Properties Arising at This Level of Structure
Organ system	Nervous system	Connection of and communication between organs all involved in a similar process can coordinate activities effectively (such as proper progression of food through the mouth, pharynx, esophagus, stomach, and intestines).
Organism	Human	All the organ systems together are required to create an autonomous unit that isn't physically and permanently connected to other units.
Population	A group of humans living in a town	Groups of individuals may form the capacity for sexual reproduction, which is a powerful evolutionary tool.
Community	The humans and other animals that live in an alpine town	Having plants and animals in a localized geographical region allows interflow of energy and organic molecules for the first example of a self-perpetuating system.
Ecosystem	All the organisms and environmental factors in an alpine town	Local environments vary in temperature, rainfall, terrain, and other abiotic factors that shape the organisms within it through natural selection. Organisms are a part of this because they, too, exert shaping forces on their neighbours.
Biosphere	Earth	Different ecosystems are not isolated, so this level of complexity encompasses the incredibly complex influences different ecosystems have on each other as well as large-scale changes that occur to all of them at once (such as climate change).

CHAPTER 7 STUDY STRATEGIES

B

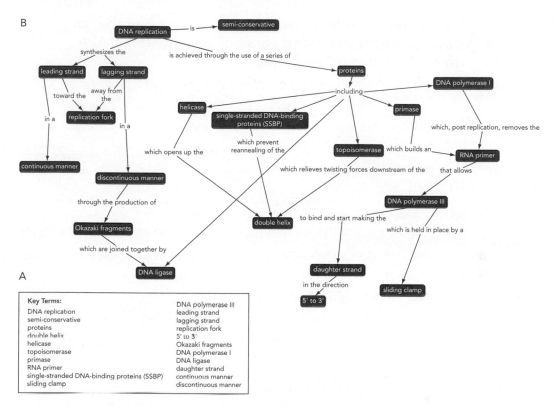

Figure 7.1   (a) A list of key terms associated with the process of DNA replication. (b) A concept map created based on the key terms in (a).

CHAPTER 8 SCIENCE AND COMMUNICATION

Table 8.1 Examples of APA Reference Styles for a Number of Source Types. To see an extended version, along with an APA tipsheet, go to the *Science*[3] website.

BOOK				
Author.	(Date).	*Title of book.*	Location:	Publisher.
Kandel, E. R. (2012). *The age of insight: The quest to understand the unconscious in art, mind, and brain, from Vienna 1900 to the present.* New York, NY: Random House.				

PERIODICAL (PRINT OR ONLINE)						
Author.	(Date).	Title of article.	*Title of Periodical,*	Volume(Issue),	Page(s).	DOI (if available)
Iezzoni, L. I., & Ogg, M. (2012). Patient's perspective: Hard lessons from a long hospital stay. *American Journal of Nursing, 112*(4), 39–42. http://dx.doi.org/10.1097/01.NAJ.0000413457.53110.3a						

PERIODICAL (NO AUTHOR)					
Title of article.	(Date).	*Title of Periodical,*	Volume(Issue),	Page(s).	DOI (if available)
Borderviews: Light plain/plane of light. (2012, March/April/May). *Border Crossings, 31*(1), 18.					

ONLINE/DATABASE ARTICLE (NO DOI)						
Author.	(Date).	Title of article.	*Title of Periodical,*	Volume(Issue),	Page(s).	Retrieval Statement
Jennings, H., Nedeljkovic, M., & Moulding, R. (2011). The influence of confidence in memory on checking behaviours. *E-Journal of Applied Psychology, 7*(2), 9–16. Retrieved from http://ojs.lib.swin.edu.au/index.php/ejap/article/view/192/278						

ARTICLE AVAILABLE ONLY IN A DATABASE (NO DOI)						
Author.	(Date).	Title of article.	*Title of Periodical,*	Volume(Issue),	Page(s).	Retrieval Statement
Thompson, K. J., Thach, E. C., & Morelli, M. (2010). Implementing ethical leadership: Current challenges and solutions. *Insights to a Changing World Journal,* (4), 107–130. Retrieved from http://www.franklinpublishing.net/insightstoachangingworld.html						

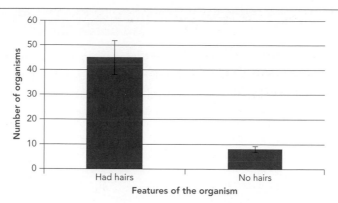

Figure 1 This purely hypothetical bar graph demonstrates how to show data in a bar graph. Note that this informative caption occurs underneath the figure to which it refers. A bar graph is appropriate because the organism either had hairs or did not have hairs. Error bars show variation between different measurements.

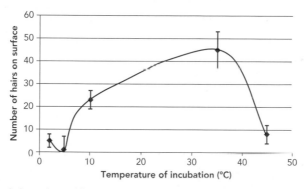

Figure 2 This purely hypothetical line graph demonstrates how to show data in a line graph. Note this other informative caption. A line graph is appropriate because the data vary over a range of temperatures and you can suppose that an intermediate number of hairs will grow if the organism is incubated at 22°C.

Table 1 This purely hypothetical table demonstrates how to show data. Note that this informative caption appears above the data presented, unlike in figures.

Data Point	Value
1	45.3 ± 3.1
2	21.4 ± 9.6
3	15.8 ± 3.3

Figure 8.2 A figure of figures (and a table). Note that all have captions that give enough context to the viewer so that it could stand alone. Your manuscript will refer to the image by name (Figure 1 or Table 1, as appropriate). Captions go underneath figures and above tables. The numbering system is independent for figures and tables (the number increases with each figure, starting at 1 from the first figure shown; the number increases with each table, starting at 1 from the first table).

A Trick: Don't Write in Order!

Because you've done the work already and it's fresh in your mind, write the Method section FIRST, then proceed through the other things you've done. Here's a suggestion for the order in which you should do the actual writing (just remember to put them in the proper order before submitting!).

- Method

- Results

- Discussion and Introduction (You decide which is easier to start with, but you will have to tinker with these alternately to make sure they form two parts of the same thing! For example, you can refine your hypothesis a bit to accommodate changes you made in your experiment and make the implications of your results more clear.)

- References

- Abstract (Referring to the already-written manuscript for this summary makes it easier!)

- Title (Strangely enough, this is easiest at the end! You'll have all the important ideas sorted out and can make a descriptive, complete title with little effort!)

CHAPTER 9 JOB, SCHOLARSHIP, AND POST-GRADUATE APPLICATIONS

BOX 9.1 **Transferable Skills in Science**

Scientific Skill	Place You Learned It	How the Skill Transfers
General science skills	Science courses—lecture, lab, and exams	Appreciate the complexity of the Universe and of life.
		Think critically to evaluate data and evidence.
		Have a toolbox full of methods for studying a new topic.

Continued

Scientific writing	Literature reviews Lab reports Lab notebook Mini-essays (1–5 pages) Major papers Poster presentations Tests/exams	Communicate in a variety of forms. Know appropriate citation and referencing methods. Give clear and accurate accounts of a topic in writing.
Oral communication	Solo presentations Group presentations Discussions Debates	Give a clear and accurate verbal report. Facilitate a discussion.
Numeracy	Collecting data Analyzing data Interpreting data Representing data in graphs Statistical analysis Mathematical modelling Evaluating conclusions derived from data	Understand sampling and sampling errors, calibration, and types of error. Obtain appropriate data. Interpret and present data. Perform statistical analysis of data. Use data in various forms.
Computer skills	Exposure to, and use of, a variety of software programs	Use computers and various types of software to solve problems and work with data. Use computers to do literature searches. Use computers and diverse types of software to communicate.
Experimental design	Developing hypotheses Performing hypothesis testing Designing methods Following methods Analyzing experimental outcomes	Design, plan, and carry out an experiment. Evaluate results and modify the experiment.

Continued

Lab skills	Employing general lab techniques Observing experiments Recording data	Be comfortable in a lab setting. Learn new techniques as needed. Keep a record of work done.
Field skills	Observing in the field Collecting specimens Experimenting Identifying organisms in natural habitats	Work in the field solo or as part of a team. Obtain and analyze data. Follow legal and ethical requirements.
Proficiency with scientific equipment	Using various instruments in labs	Work with common equipment in your field. Learn to use specialized equipment.
Literature search and information retrieval	Completing background research for lab reports Studying for exams	Read and evaluate literature critically. Discern between poor, good, better, and best sources of information.
Literature critique	Critical reading Critical reasoning Assessing the literature Evaluating the literature	Discern between poor, good, better, and best sources of information. Analyze and summarize information.
Problem solving	Completing problem sets Participating in group work Doing "real" labs (i.e., non-cookbook labs)	Apply knowledge to solve problems individually or as part of a team.
Decision making	Participating in lab sessions Participating in group projects	Work alone or with a group to set goals and make decisions.
Leadership	Leading group projects Leading group work in class	Take the lead in a group while respecting other peoples' opinions. Keep others on task. Work with people from other backgrounds.

Continued

Teamwork accountability	Participating in group projects	Work as part of a team and take responsibility for your part.
		Evaluate your work as part of group.
Organizational skills	Obtaining a degree	Work independently.
	Completing assignments on time	Manage your time effectively.
	Studying for exams in a timely fashion	Set goals and work towards achieving them.
Ethics	Taking an ethics class or discussing ethics as part of a class	Identify, communicate, and debate ethical issues in science.

Source: QAA, Subject benchmark statement—Biosciences 2007. Retrieved from http://www.qaa.ac.uk/Publications/InformationAndGuidance/Pages/Subject-benchmark-statement-Biosciences.aspx, 2007; A. Jones, R. Reed, and J. Weyers (2012). *Practical Skills in Biology* (5th ed.). Toronto, ON: Pearson Education.

BOX 9.2 Sample Cover Letter

15 Waterloo Road
Metropolitan, Saskatchewan
S0S 0S0
4 July 2012

Mr. Walter Langorski
Human Resources Manager
Astrobe Laboratories Ltd.
1632 Bayonne Street
Two Planes, Ontario
M0M 0M0

Dear Mr. Langorski:

I am writing in application for the position of Junior Technician with Astrobe Laboratories Ltd. I think the water-quality tests you perform for industry are very

Continued

innovative, and I would love to have the opportunity to be associated with such a forward-looking company. I have enclosed my resumé for your consideration.

Through my studies in the Environmental Chemistry program at South Atlantic University, I have developed a number of lab skills that I believe will be useful to your company. These include Soxhlet extractions, the usual distillation and titration techniques, and lots of experience in structure determination of unknown compounds. I have also had the opportunity to work with instrumentation, including both hydrogen and carbon NMR, ICP-MS, GC-MS, Raman, FTIR, and UV-visible spectroscopy. Writing weekly laboratory reports has helped me improve my communication skills.

As the team leader for a successful undergraduate research project, I developed organizational skills while ensuring everyone in the group had their work done by the required deadlines and preparing our poster for the Student Research Day at South Atlantic University. I am really good at both following instructions and being a self-motivated worker, as the circumstances require.

I would like to apply my laboratory and leadership skills to this position with Astrobe Laboratories Ltd. I hope you will give me an opportunity to meet you in an interview. Thank you for reviewing my application.

Sincerely,

James Hurdsman

James Hurdsman

BOX 9.3   Cover Letter Checklist

- Clearly state the position that you are applying for.
- Show how you are qualified for the position (i.e., draw the reader's attention to the aspects of your resumé that make you the person they want).
- Show that you understand what the position entails.
- Show that you are knowledgeable about the company/program/department— go on their website and do some research before you start writing.
- Show how you will fit into their departmental or corporate culture.

Continued

- Show how you will be a benefit to the company, department, or program.
- Show that you can communicate clearly and well.
- Show that you are mature and capable, but don't come across as arrogant!

BOX 9.5 **Sample Personal Statement in Response to the Question "Explain Why You Want to be Head of the Laboratory Safety Department in 500 Words or Fewer"**

I am applying for the position of Head of the Laboratory Safety Department because I believe I would be a good asset to the department and a strong leader. I have enjoyed the challenges as Deputy Head this year and believe that my strong administrative background helps me be effective in the job. I have previously served as Biology Safety Coordinator and worked closely with the previous department head, so I was familiar with many aspects of the job when I stepped into it. I have enjoyed the increased opportunities to lead the department in a very exciting time as we deliver, for the first time, a new program of laboratory safety at this institution.

The key strengths that I possess for success in this position are a great liking for the people I work with, deep integrity, and a strong work ethic. I think I have shown over many years at this institution that I am an innovative thinker and good at motivating others to move forward, even when confronted by unknown territory. A project I have greatly enjoyed working on is designing the new Hazardous Wastes Disposal Unit to serve the needs of this institution well into the future. I was instrumental in developing a new organizational system for assessing risk and obtaining funding from the government of Hidalgo for new instrumentation to track hazardous vapours to their source, and have encouraged my colleagues to develop waste-disposal tracking sheets. In addition, I have involved our colleagues throughout the institution by pioneering the idea of waste-disposal systems that could be used by researchers from various disciplines. I am excited by the new Safety Assessment Seminar that we have developed and are piloting for the first time this month.

I always try to lead by example, never requiring someone to do something that I would not or am not already doing myself. I believe that our department often

Continued

leads the way in embracing change, and I feel strongly that as head I would continue to encourage this in a well-thought-out and mutually-agreed-upon fashion. I believe arriving at consensus in departmental decisions is important and plan to continue to build on the foundation laid by our previous head in this area. I know that I will sometimes make, and indeed have made, mistakes, but I am willing to acknowledge the mistakes I make and learn from them.

I trust that my dedication to the Laboratory Safety Department and this institution is evident to you and to the committee. I look forward to the opportunity to discuss my application with the selection committee.

BOX 9.6 Sample Resumé

Yulissa MARKISON   805-216-5034

3-81 Line Way, Whalleye, Ontario L1T OS9   ymarkison@jiffymail.com

OBJECTIVE: APPLIED SCIENCES LABORATORY TECHNOLOGIST
SUMMARY:

- Multidisciplinary knowledge of biochemistry, microbiology, and molecular biology obtained through both college and university education and work experience.

- Prepared and calibrated standard samples. Operated the instruments independently.

- Working knowledge of techniques for cell biology, protein purification.

- Experience in collecting, analyzing, and interpreting data. Experience keeping a formal laboratory notebook following SOP and USP guidelines.

- Effective interpersonal, organizational, and project management skills developed through professional experience.

- Working knowledge of various data acquisition software applications and Microsoft Office.

- Total time spent in both college and university laboratories: over 700 hours.

Continued

PROFESSIONAL SKILLS:

Chemistry:
Solution preparations, titrations, and gravimetric analysis. Product content analysis, as well as instrumentation techniques such as spectrophotometry (UV/Vis, IR, NMR), chromatography, flame emission, atomic absorption, HPLC, and GC.

Biochemistry:
Preparation of agarose gels and SDS–PAGE gels from scratch, protein purification by chromatographic techniques (size exclusion, ion exchange, affinity), DNA extraction and PCR, protein assay/characterization, SDS-PAGE, Western blotting, ELISA.

Microbiology:
Tissue culture aseptic techniques, cell passage, media preparation, dilutions, plate pouring, plate counts, spore transfers, plate streaking, gram staining. Environmental analysis as well as product analysis following USP guidelines.

Statistical Methods in Quality:
Knowledge of bioinformatics and modelling software, GLPs, GMP, WHMIS, IUPAC, BIOSAFETY, and other regulations. Authored reports and summaries, performed data entry.

Technical Communications:
Writing technical reports, standard operating procedures, certificates of analysis, formal laboratory notebooks for all disciplines of study.

WORK-RELATED EXPERIENCE:

Laboratory Technologist Assistant   April 20XX–September 20XX

Eastern University

- Data entry
- MSDS project management
- Inventory management
- Supply management
- Shipping and receiving
- Interpreted data
- Wrote technical reports

Continued

Quality Inspector   April 20XX–September 20XX

The Inspection Group

- Inspected manufactured parts to ensure quality standards are met.
- Managed all quality-related documentation.
- Oversaw staff schedules, balanced overtime, and submitted attendance records for payroll.
- Supervised a team, acted as a liaison between quality assurance staff and the project manager.
- Participated on internal auditing team to maintain the quality standards of an ISO-9000 certified company.

EDUCATION:

B.Sc. Cell and Molecular Biology   September 20XX–April 20XX

Eastern University

Biotechnology Technologist   September 20XX–April 20XX

Northern College

3-year Ontario College Advanced Diploma

Graduated with Honours

Graduated on Dean's List

References are available upon request.

Box 9.7 Sample Scannable Resumé

Yulissa MARKISON

805-216-5034

3-81 Line Way

Whalleye, Ontario L1T OS9

ymarkison@jiffymail.com

Continued

OBJECTIVE: APPLIED SCIENCES LABORATORY TECHNOLOGIST

SUMMARY:

- Multidisciplinary knowledge of biochemistry, microbiology, and molecular biology obtained through both college and university education and work experience.

- Prepared and calibrated standard samples. Operated the instruments independently.

- Working knowledge of techniques for cell biology, protein purification.

- Experience in collecting, analyzing, and interpreting data. Experience keeping a formal laboratory notebook following SOP and USP guidelines.

- Effective interpersonal, organizational, and project management skills developed through professional experience.

- Working knowledge of various data acquisition software applications and Microsoft Office.

- Total time spent in both college and university laboratories: over 700 hours.

PROFESSIONAL SKILLS:

Chemistry:
Solution preparations, titrations, and gravimetric analysis. Product content analysis, as well as instrumentation techniques such as spectrophotometry (UV/Vis, IR, NMR), chromatography, flame emission, atomic absorption, HPLC, and GC.

Biochemistry:
Preparation of agarose gels and SDS–PAGE gels from scratch, protein purification by chromatographic techniques (size exclusion, ion exchange, affinity), DNA extraction and PCR, protein assay/characterization, SDS-PAGE, Western blotting, ELISA.

Microbiology:
Tissue culture aseptic techniques, cell passage, media preparation, dilutions, plate pouring, plate counts, spore transfers, plate streaking, gram staining. Environmental analysis as well as product analysis following USP guidelines.

Statistical Methods in Quality:
Knowledge of bioinformatics and modelling software, GLPs, GMP, WHMIS, IUPAC, BIOSAFETY, and other regulations. Authored reports and summaries, performed data entry.

Continued

Technical Communications:
Writing technical reports, standard operating procedures, certificates of analysis, formal laboratory notebooks for all disciplines of study.

WORK-RELATED EXPERIENCE:

Laboratory Technologist Assistant

April 20XX–September 20XX

Eastern University

- Data entry

- MSDS project management

- Inventory management

- Supply management

- Shipping and receiving

- Interpreted data

- Wrote technical reports

Quality Inspector

April 20XX–September 20XX

The Inspection Group

- Inspected manufactured parts to ensure quality standards are met.

- Managed all quality-related documentation.

- Oversaw staff schedules, balanced overtime, and submitted attendance records for payroll.

- Supervised a team, acted as a liaison between quality assurance staff and the project manager.

- Participated on internal auditing team to maintain the quality standards of an ISO-9000 certified company.

EDUCATION:

B.Sc. Cell and Molecular Biology

September 20XX–April 20XX

Eastern University

Continued

Biotechnology Technologist

September 20XX–April 20XX

Northern College

3-year Ontario College Advanced Diploma

Graduated with Honours

Graduated on Dean's List

References are available upon request.

BOX 9.9   Sample Thank-You Letter

Mr. John Smith
111 Anywhere St.
Lane's End, Alberta, T0S 0K0
Canada
29 August 2012

Rebecca McGee
654 Mount Albert Road
Knoxville, Alberta, T1V 9V9
Canada

Dear Mr. Smith:

Thank you so much for the Smith Chemistry Award, which I received this year. It was such an honour, and I really appreciate the effort that your family takes to ensure that students each year will be aided financially in their studies.

This award will help me to purchase textbooks this year without having to take on a second part-time job. It was also really encouraging to know that someone felt that I have the ability to succeed in chemistry. Sometimes in the middle of the semester it is easy to lose sight of my goals, and this award supplied a great emotional lift as well as the more obvious financial help.

Continued

Since I really enjoy chemistry and hope to go on to graduate school once I have finished at South Atlantic University, the Smith Chemistry Award has provided added incentive to work toward my goals. As a first-generation Canadian, I have had to climb a lot of barriers to enter university, and I am very grateful for this award. Thank you again for the Smith Chemistry Award.

Sincerely,

Rebecca McGee

Rebecca McGee

BOX 9.10	Special Examinations Required for Application to School

Name of Exam	What It Is Used For	Where to Get More Information
MCAT (Medical College Admission Tests)	Part of the application for all U.S. and most Canadian medical schools.	www.aamc.org/students/applying/mcat/ (This is the official website and has information for both U.S. and Canadian students.)
DAT (Dental Aptitude Test)	Used by Canadian dental schools to assess applicants. A similar test, the Dental Admission Test, is used by U.S. schools.	www.cda-adc.ca/en/dental_profession/dat/ (This site is hosted by the Canadian Dental Association.) www.ada.org/dat.aspx/ (This is the link for the American Dental Association test.)
GRE (Graduate Record Examination)	An admission requirement for many graduate schools in the United States.	www.ets.org/gre/ (This is the website for the group that owns this test.)
OAT (Optical Admission Test)	An admission requirement for all schools of optometry in the United States. No information was found for the two schools in Canada.	https://www.ada.org/oat/index.html (This is the website for the group that owns this test.)

Continued

PCAT (Pharmacy College Admission Test)	An admission requirement for pharmacy schools in North America.	pcatweb.info/ (This is the website for the group that owns this test.)
VCAT (Veterinary College Admission Test)	No longer used. Most veterinary schools now require MCAT or have no particular admission exam requirement.	www.aavmc.org/Publications/VMSAR.aspx (This is a link to a publication regarding veterinary school admission requirements in the United States and Canada.)
SAT Reasoning Test (formerly known as Scholastic Aptitude Test)	Entry to most universities in the United States (competes with ACT).	student.collegeboard.org (This is the website for the group that owns this test.)
ACT (formerly known as American College Testing)	Entry to most universities in the United States (competes with SAT).	www.actstudent.org/ (This is the website for the group that owns this test.)

BOX 9.11 Possible Sources of Funding for Science Research in Canada

Name of Funding Agency	Contact Information
Natural Sciences and Engineering Research Council of Canada (NSERC)	www.nserc-crsng.gc.ca
Canadian Institutes of Health Research (CIHR)	www.cihr-irsc.gc.ca
Social Sciences and Humanities Research Council of Canada (SSHRC)	www.sshrc-crsh.gc.ca
National Research Council of Canada (NRC)	www.nrc-cnrc.gc.ca
Research, Technology and Innovation (Industry Canada)	www.ic.gc.ca
The Canadian Institute for Advanced Research (CIFAR)	www.cifar.ca
Canadian Innovation Centre	innovationcentre.ca

Continued

Canada Foundation for Innovation (CFI)	innovation.ca
InNOVAcorp	innovacorp.ca
International Opportunity Fund (IOF)	www.igfagcr.org/iof-home-page
Canadian Water Network (CWN)	www.cwn-rce.ca

Box 9.14   Sample Thank-You Letter

15 Waterloo Road
Metropolitan, Saskatchewan
S0S 0S0

1 August 2012

Mr. Walter Langorski
Human Resources Manager
Astrobe Laboratories Ltd.
1632 Bayonne Street
Two Planes, Ontario
M0M 0M0

Dear Mr. Langorski:

Thank you very much for the opportunity to interview for the position of Junior Technician with Astrobe Laboratories Ltd. It was really nice to meet you and your colleagues, and I particularly enjoyed the tour of the facility.

Since the interview I have been even more excited about the possibility of working for your company, and I hope that you will consider my application package and interview in a positive manner.

I would like to thank you once again for the opportunity to meet with you, and I hope to hear from you soon.

Sincerely,

James Hurdsman

James Hurdsman

Index

science, as web of connections, 49
scientific method, 49
observing the universe, 55–56
Secondary source, 358
Second law of thermodynamics, 198
Significance level, 160
SI (Système International) units, 179, 388
prefixes, 385
Soddy, Frederick, 207
Solid state physics, 193
Species named after, 106–107
acronym, 107
celebrities, 107
commercial products, 106–107
misnamed, 107
taxonomy, 106
on their location, 107
varying reasons for, 106
Statistics and fitting, 149–157
averages, 152–155
mean, 154–155
median, 154–155
mode, 154–155
distribution, 150–152
frequency distribution, 152
probability distribution, 152
sampling, for population, 158–162
variables, 149–150
interval, 150
nominal, 150
ordinal, 150
variation, 155–158
standard deviation, 155–158
variance, 155–158
Study strategies, 245–261
after every class, 252–253
condensing notes, 255
before every class, 248–250
during every class, 250–252
key, to keep in mind, 261
labs and tutorials, 259
learning outcomes, 248
practice tests, 255–257

real test, writing, 258
study schedule, 254
syllabus of course, 246–247
time management, 259–260
use of professor office hours, 253
Substitution or single-displacement reaction, 215
Sufficiency, 28
Suffix, 84
Summarizing, 278–279
Symbols, in science, 80, 363–364
Synthesis or combination reaction, 215
Systematics, 227

T

Taxonomy, 106
Tertiary source, 358
Tesla, Nikola, 186
Test yourself, 355
Textbook, use of, 248
Thank-you letters, 417–418, 420
Theoretical chemistry, 212
Theory, 61
Thermodynamics and statistical mechanics, 196
Tinbergen, Nikolaas, 221
Transferable skills, in science, 312–315
True understanding, 355

U

Uncertainty, 23

V

Van't Hoff equation for osmotic pressure, 214
Variation, 155–158
standard deviation, 155–158
variance, 155–158
Vector, 140
conventions and operations, 141
addition, 141
cross product, 141
dot product, 141
magnitude, 141
multiplication, 141

INDEX